A Most Industrial Life

– and its lessons for our future

Also by Philip Adams:

A Most Industrious Town: Briton Ferry and its people, 1814–2014

Not in Our Name: War dissent in a Welsh town

Daring to Defy: Port Talbot's war resistance 1914–1918

Re-shaping Rail in south Wales: the railways of Briton Ferry and district – past, present and future
(with Martin Davies)

Bonjour!
Ludlow French Twinning: a unique experience

A MOST INDUSTRIAL LIFE

– and its lessons for our future

Philip Adams

Ludlow and Neath-Port Talbot 2024

Published by Philip Adams
144 Corve Street, Ludlow, Shropshire, SY8 2PG, UK
www.britonferrybooks.uk

First published 2024

Text copyright © Philip Adams 2024
The right of Philip Adams to be identified as the author of this work has been asserted by him in accordance with the Copyright, Designs and Patents Act 1988

All rights reserved. No part of this publication may be reproduced, stored in a retrieval system, or transmitted in any form or transmitted by any means, electronic, electrostatic, magnetic tape, mechanical, photocopying, recording, or otherwise, without prior permission of the author of the works herein.

ISBN 978-0-9930671-5-0

Typeset in Times and Frutiger and designed by Internet@TSP, Ludlow
Printed and bound in UK by Flexpress Ltd, Leicester

Dedication

To everyone mentioned in this book, and to unnamed others in 'My Industrial Life', whether in a personal or an occupational context, who were positively involved as witness or participant at the times and places mentioned.

Author's note

The chronological, diary style, presentation of *A Most Industrial Life* enables readers to approach the whole work from the beginning, or to select time-periods of particular interest. It is based on an assemblage of personal experience and general events which have affected the author both directly and indirectly.

Whether narrating as a witness or participant, the author has focussed on many of the significant events which culminated in the deindustrialisation of Britain. Some typical cases are featured separately:
- 'The UK Car Industry in the 1960s and 1970s' – page 43
- 'Joyce and Derek Ridings' – page 129
- 'Decarbonising or Deindustrialisation – Port Talbot 2024' – page 168

In particular through the example of steelmaking at Port Talbot it outlines the sort of change that is needed to recover.

Restructuring of policy and practices, by Government, the private sector and ourselves, will be necessary to achieve the decarbonisation required in that recovery. Changes in both working and daily life are also essential ingredients.

People are mentioned from time to time to provide an authentic narrative. Some names are in the public domain, but those specifically mentioned do not comprise a comprehensive record of everyone involved in this story.

Introduction

Briton Ferry, the town of my birth, lies in the county borough of Neath Port Talbot. It was a most industrious town. There is absolutely no doubt about that. Yet it was much more than an industrious town: it lay in a beautiful setting between the woodlands and the sea, towards which flowed the river Neath, on whose banks much of the town's prosperity was created. That was why the likes of Turner and Horner wanted to capture the original Briton Ferry on canvas. And did so, so well.

Ruskin Street, the street where I once lived, was named after John Ruskin, the polymath. As a pioneering conservationist, who foresaw the 'greenhouse' effect more than a century ago, Ruskin inspired the establishment of the National Trust, and the founders of the National Parks movement. His doctrine was that work should be meaningful and should improve workers' wellbeing. Workers should have dignity, show citizenship and a sense of community. Once, this was the case in Briton Ferry, the town of my birth and in the street where I once lived.

The developing town was also a community. What was lost through industrialisation was gained from a social spirit engendered during the industrialisation process. It was deindustrialisation that took me away from that town to others. There, on reflection, what had happened in Briton Ferry seemed to happen again and again and is still happening today. It was as though one was running to stay still. It is time to question what was really happening over this period of fifty years, and why? Many of the events described in this work have directed me to an answer.

Deindustrialisation, and of manufacturing especially, is not the only theme that dominates 'My Industrial Life'. Despite the shift from making things to service activities, environmental matters are prominent, too, in the references to pollution and man-made climate change. Whilst most pollution may be industrial in origin, the greenhouse gas emissions that are causing climatic disaster derive from both everyday life and industrial activities.

Therefore, this work weaves personal events into an historical and industrial context. It is a selective account, but which should nevertheless have relevance and meaning for you, even though you may not have been around at many of the times and places mentioned.

**Ruskin Street (2010) –
named after John Ruskin**

1940s

I have few recollections from the 1940s regarding our home in Ruskin Street. It was in an Edwardian end-but-one, three-bedroomed terraced house, brick built with a slate roof and sash windows. They rattled. In wintertime, before the advent of central heating, ice would often form inside the glazing, but we occupants used hot water bottles. There was a longish rear garden with an outside WC and coalhouse at the bottom. There, a common lane was shared with neighbouring Glan-y-mor Street which ran parallel. For some time, an Anderson shelter and even a chicken-run were in place in the back garden. These were both hangovers from World War Two. Each garden was separated by a brick wall about 1.75 metres high, making it easy enough to speak with our neighbours, all of whom we knew. Later houses were semi-detached and in blocks of four. They had front gardens with a plenitude of privet hedges.

We walked to the primary school, played in the street because there were few vehicles and knew all our neighbours.

Heating for the house was from a black, coal-fired range which incorporated a small oven which also provided some hot water from a back boiler. We had a bathroom, but when small we often bathed in a zinc bath in front of the range. Coal for the range was kept in an outhouse at the bottom of the garden-opposite a lavatory which was not replaced with an inside toilet until the 1960s.

One memory is a fire which destroyed the Lodge Garage in Briton Ferry. I also remember my grandfather, Jim Adams, who died when I was five. He kept railway books in a tall cupboard on the side of the fireplace in our living room. It would be another 60 years and more before the school closed and I found out the significance of the railway books after research into my family's history. Our shed had his taps, dies and drills, and shoemaker's lasts to repair our shoes, as well as a mysterious medal.

Even as young children, we would wander the locality to play all day in the woods, the park, or the canal and explore the railways and riverside wharves dotted around the area, often to retrieve trophies from the scrap, such as steel helmets left from the war.

Brunel hydraulic tower BF dock (Martin B Davies)

Little did we children know that we were sometimes playing on historic industrial structures like Brunel's hydraulic tower on the dockside whose monuments are a Grade II listed area of national importance. The main industrial archaeological interest is the innovative use of hydraulically operated buoyancy chambers in the lock gate.

Behind these recollections lurked the 1942 *Report on Social Insurance and Allied Services*. It advocated a cradle-to-grave social policy to combat the five giants of idleness, ignorance, disease, squalor and want that still aggravated the country. It became the blueprint for social policy in post-war Britain which influenced Prime Minister Attlee to introduce the National Health Service Act of 1946 which created the NHS in July 1948.

1950s

In 1950 I started at Brynhyfryd School. Life was spent with my parents and two brothers in Ruskin Street, as well as my father's elder brother, John Adams. My father worked as an insurance agent in the Neath area, but his office was in Port Talbot. My mother was very well respected both for her skills and knowledge as a retired nursing sister and for her opinions.

Uncle John's story was unknown to we brothers at that time and, once again, was not fully revealed for another sixty plus years until my second book *Not in Our Name* spilt the beans: he was in prison during World War 1 with G M Ll Davies, friend of Lloyd George, and James

The Adamses at Brunel's bridge in Chepstow

The 'school' park – River Neath and now-demolished BP oil refinery

Maxton among others. My first book (*A Most Industrious Town*) had touched on his background and included a short, personal, chapter entitled *Peace time memories of life on a Briton Ferry Street* which depicted life for the children who lived there at that time.

I remember the parquet floors of Brynhyfryd Primary School assembly hall for several reasons. It was whilst sitting on the floor that Headmaster Henry Trick, once a naval person, claimed that 'the sun never set over the British Empire'. It was there, too, that teacher 'Tosh' Evans once entered with such gravitas to announce that 'the King is dead'. Each week, Mr Trick ceremoniously updated the board of honour to show how each house was performing in the school league table for 'cleanliness, conduct and punctuality'. He also brought theatre to the school and had a stage built in that very same hall. It was there, too, that we sang Welsh songs and hymns and took part in Eisteddfodau.

The three Adams boys were together until 1963

The mid-1950s brought the 11-plus exams to the three Adams boys of Ruskin Street. In quick succession we transferred to the Grammar School in nearby Neath. Comprehensive education was still some way off. I was encouraged to take up sport, although I had shown no sign of sporting prowess, or even interest whilst in Brynhyfryd Primary School. Yet I loved going with my father to watch Briton Ferry rugby play at Cwrt Sart and with the spare match ball, try to kick goals. When he said that it was time to go, I said that I wanted to stay.

13

Two of the sporting events that remain in our memories were the Empire Games in Cardiff and the football World Cup held in Sweden in 1958. The glory days of Welsh Rugby were yet to come. As early teenagers we were aware of the Munich air crash, also in 1958, when twenty-three of the forty-four on board the flight from Belgrade to Manchester died. After two abandoned take-offs, a third resulted in a runway over-run due to slush on the runway.

One important characteristic of the 1950s was the solidarity of the street in which we lived. It was a community. For example, when television started to become available in the early 1950s it was not unusual for Saturday night entertainment to take the form of watching black and white television in the home of a neighbour who could afford to rent or buy a TV. Car ownership was rare and a second feature of life in the 1950s was that neighbours would join for a bus trip, perhaps to Tenby, Chepstow, or Symond's Yat.

Reflecting widely on the 1950s, the decade saw not just the end of Attlee's Labour government in 1951, but also the gradual ending of Churchill and Eden's supervision of the British Empire. In South Africa in 1960, Prime Minister Macmillan recognised that Labour's decolonisation process must be continued. The British Government could not sustain the British Empire and the 'winds of change' had to be allowed, to avoid Soviet penetration into Africa and elsewhere.

Riverside wharves and park

1960s

Somehow, I found myself in the under-15 rugby teams for both the grammar school and for Neath schoolboys. I cannot remember whether I got involved in rugby before or after being conscripted into athletics. And I cannot understand why I was conscripted into either, or by whom. I know it was either by Roy Bish or Ron Trimnel, PE teachers at the school. Bish moved on, via Cardiff teacher training college, to become first coach of the Italian national rugby team. His influence put Italy on the road to joining the Six Nations Championship.

Long after a winter trip to play for Neath at Chingford, Essex, it suddenly dawned on me who the family were that hosted us. I am uncertain whether it was Mark or John Tebbit who played against us that snowy morning, but their father did not watch the game. He wished us the best of luck and set off to work. Years later I realised the significance of his short-sleeved white shirt: a pilot's shirt,

Neath Grammar School Under-15 Rugby Union team

and the wearer was none other than Norman Beresford Tebbit, then of BOAC–British Overseas Airways, now Lord Tebbitt of 'get on yer bike' fame.

No-one was more surprised than me that I became a final trialist for Wales under-15 schools as a wiry loose head prop-forward. My father said nothing then nor since, but Albert O'Shea, a friend of his on Briton Ferry RFC's committee told me, long after my father's death, that the cap I might have received went to a selector's son!

That year was also my first foreign visit: a school trip to Menton by Channel ferry and steam train, where one highlight was to see William Webb-Ellis' grave in *le cimetière du vieux château*. It had recently been rediscovered by the co-founder of the Guinness Book of Records, Ross McWhirter, and is now maintained by the French Rugby Federation.

After that I played a few games for the Grammar School's second team, but I became disillusioned with its attitude to those who favoured association football. The school seemed to want to protect its impressive rugby record at all costs, but I thought that its illiberal stance was too high a price to pay. Consequently, I focussed much more on track and field athletics and cross-country running. I also played for local soccer and rugby teams and thereby keeping some friends in those teams from primary school, but who had not attended Grammar School. In the under-14 group, I first ran 59.2 secs for 440 yards in Neath area schools' sports on a bumpy grass track in plimsoles. It was fear that kept me going. Gradual improvement took me to run 49.3 within a decade, winning the 1964 junior title at the Welsh Games.

Little has passed my lips about academic subjects, but there was an ill-advised separation between arts and science subjects after the second year in grammar school, without any effective guidance about the consequences of the path of study chosen. The separation did not just apply at that school; it was a national problem. The imbalance in the curriculum between arts and science does not apply to other west European countries.

The early 60s was the time of the Cold War and the Cuban missile crisis. At the initiative of a forward-thinking teacher, Russian was taught, using BBC resources, but it was too much for the traditional, autocratic headmaster who soon put a stop to it.

By the early 1960s, teenagers were already different to those of the previous decade. The first teenage generation, free from conscription, emerged in Britain. Young people were finally given a voice and freedom to do what they wanted. The parents of the Sixties teenage generation had spent their youth fighting for their lives in the War and wanted their own children to enjoy a youth of more fun and freedom.

Technological advancements of the 1960s changed how people spent their leisure time. Increased income through employment in factories allowed people to spend more on leisure activities.

Television and pocket transistor radios allowed people to spend their free time listening to music and watching TV. White goods shortened the amount of time people spent on housework, allowing them more freedom and time to enjoy themselves. The first men were on the moon in 1969, to end the decade on a note of optimism and to be able to dream for something better.

By the start of the 1960s, after 110 years, ironmaking had ceased at Briton Ferry. All that remains today of the riverside works is the 1910 engine house which had produced the power to provide hot blast from the stoves to the furnace. It was Grade 1 listed in 2000.

The tinplate and steel sheet works in the town suffered a similar fate, with none remaining in production by 1960, having been gradually consolidated into the new Steel Company of Wales operations at Port Talbot, Trostre and Velindre during the 1950s.

1963

I was a proud team captain and head prefect at the Grammar School in 1963–64 when an initially naive team improved so much that we went on to win 11 of the 14 cross-country races run. Alongside that I somehow contrived some, unexpected, good, A-level examination results.

A visit with school to Lido di Jesolo on the Venetian Riviera and Postojna caves in 1963 was my second overseas trip. School friend John Blackmore suggested that we hitch-hike to Aberdeen, and to visit his new girlfriend near Glasgow. We eventually got to her home after a night in a police cell, following an attempt to spend the night on benches at Glasgow Central Station. The police treated us well, but our Scottish trip ended at Inverurie. Tired of all the hitch-hiking, by the time of arrival at Warrington Bank Quay station, it was time to take a train home.

Another extraordinary trip as teenagers took us eventually to the Italian Lakes, but we had no real plan. Graham James' father kindly allowed us the use of his quite new two-tone Austin Cambridge. The exact route through France and Belgium still escapes me, but I recall having fish and chips at Blankenberge before we decided to try a route through Hunsruck in the Ardennes. No sat nav of course, and no understanding of German: I thought *Ausgang* must have been a big city indeed! I suppose that, if you are to get lost, then its best to do it properly. And that is just what we did. Somehow, we got *off piste* at a place called Idar Oberstein where we came face-to-face with a threatening American tank. Too late we learned that the town is famous for gemstones and that there is an artillery school there. Afterwards, the town erected signs which said *German Artillery Capital*. We retreated gem-free and shrapnel free.

Things continued to be eventful: the sinuous mountain passes were very demanding on the car and at times the radiator overheated. When we stopped at Bellinzona to attend to it, Graham ignored my warning not to remove the radiator cap until the system had cooled down. Although he did a very creditable backwards long jump when the radiator depressurised, he still got some nasty burns and a visit to Bellinzona Hospital for treatment.

One of Brunel's three bridges over the South Wales Mineral Railway's 1 in 10 funicular *(Martin B Davies)*

1964

It was not a surprise to receive full colours in athletics by Swansea University Athletic Board for 1964/65 and the University of Wales the following year. These awards played a part in my decision to follow fellow athlete Howard Davies and later apply for a post-graduate teaching course in education and physical education at Loughborough Colleges.

In the meantime, I worked during holiday time on building sites, as a postman, a bakery worker and as a bus conductor on South Wales Transport. Two of the bus routes I worked on were numbers 3 and 42 from Margam to St Mary's Square in Swansea. These were used by workers at Port Talbot steelworks and took one through the lower Swansea Valley, then awaiting the lower Swansea Valley Project to begin, with the aim of reclaiming the polluted land.

Over a period of about 150 years up until the 1920s, the lower valley of the river Tawe became one of the most heavily industrialised areas of the developed world. Several wealthy entrepreneurs, scientists and engineers of considerable ability were drawn to Swansea during this period to promote great innovation in the industrial processes, especially non-ferrous metals. The extent of the industrialisation that took place, at a time when there almost no environmental controls, created a legacy of chronic contamination of land and water by a great range of toxic and dangerous pollutants. Today only the 'Mond' Nickel Works remains, refining nickel for use in stainless steels, nickel alloys and the electroplating industry.

Brunel contributed to this industrialisation by building the south Wales Railway through the lower Swansea Valley. Sometimes, on the two bus routes in question, I worked with driver Glyn Williams who had a great interest in Brunel. Later he presented to many societies in the area. Soon afterwards, I would use those very buses as a passenger to get to work at the Steel Company of Wales's Abbey Works in Port Talbot.

Neath Grammar School athletics team

Welsh Games Winner's Shield – 440 yards (1964)

1965

This academic and sporting background set me up nicely to apply for a place at Swansea University. I had probably considered the University's sporting offer almost as seriously as its academic. That showed in the first year when I unwisely chose Italian as one of my three subjects, on the basis that the language would be useful for holidays, having visited Lido di Jesolo with school in 1963.

The transition from school teaching to university learning methods was difficult and I needed to repeat a year. So was the transition to being in hall at Clyne Castle. It afforded too many distractions, with The Woodman Inn in Blackpill being the principal one. The reason for exchanging Italian for the Politics and Government, which became my degree subject, was to explore new areas. Geography had been a bit tedious in year one, being a weak repeat of what I had studied at school. That was because Owen G Thomas, our school geography teacher, was as good at delivering the subject as any university lecturer.

Barry Sheerman, MP for Huddersfield, and currently Labour's oldest and longest serving MP, was a newly appointed lecturer in 1966 to Swansea University's Politics department. He covered American politics and remained at Swansea for 13 years before entering Parliament. Jack Spence was then also a lecturer in International Relations in the politics department, but his story can be left until we have reached Ludlow. Hywel Francis was a fellow student. It was he, and the south Wales Miners' Library at the University who later, as an MP, supported my published work on opposition to the

First World War in south Wales. Geraint Jenkins was another. He loved football and even came to play a game for Llansawel FC in Briton Ferry, but as a historian he was appointed Professor of Welsh History at the Aberystwyth University and then Director of the University of Wales Centre for Advanced Welsh and Celtic Studies.

University of Wales Athletics badges

1966

During my earlier working days as a student, I graced both the, then, Steel Company of Wales' Cold Mill and Water and Compressed Air Services departments as an operative. One of the tasks in the latter entailed checking water levels and taking samples of the streams feeding the steelworks reservoir at Eglwys Nunydd near Margam. And I was being paid for it! In the Cold Mill I enjoyed various tasks such as vehicle loading (fastening steel coils onto lorries for despatch). Some names of the customers still stick in my mind like Renault-Rouen and Volvo-Gothenburg. Little did I realise that some thirty-five years later I would meet with members of Renault's Works Council at Nissan to discuss the implications of the Nissan-Renault alliance for staff.

At other times, such as the afternoon of the 1966 Cup Final I worked as a feeder on the pickle line, or in its looping pit. The pickle line's job was to remove scale from hot rolled steel coil in preparation for various cold reduction processes. That 'afternoon' lasted sixteen hours – a double shift to cover for short staffing. The feeder's task was noisy and boring, but the money was good, so I was not disappointed at missing the televised final.

1966 was also the year of the Aberfan disaster, resulting from the catastrophic collapse of a colliery spoil tip. The tip had been created on a mountain slope above the village of Aberfan, near Merthyr Tydfil, and overlaid a natural spring. Heavy rain led to a build-up of water within the tip which caused it to suddenly slide downhill as a slurry, killing 116 children and 28 adults as it engulfed Pantglas Junior School and a row of houses. The tip was the responsibility of the National Coal Board (NCB), and the subsequent inquiry placed the blame for the disaster on the organisation and nine named employees. Neither the NCB nor any of its employees were prosecuted and the organisation was not fined.

Ronnie Davies, a Merthyr solicitor, with whom I played football for Briton Ferry, acted for the bereaved families' Aberfan Disaster Memorial Fund, established on the day of the disaster. It received contributions, totalling £1.75 million.

The remaining tips were removed only after a lengthy fight by Aberfan residents against resistance from the NCB and the government on the grounds of cost. The site's clearance was paid for by a government grant and a forced contribution of £150,000 taken from the memorial fund. In 1997 the British government paid back the £150,000, and in 2007 the Welsh Government donated £1.5 million to the fund and £500,000 to the Aberfan Education Charity as recompense for the money wrongly taken.

Port Talbot Steelworks

1967

A Labour government passed the Iron and Steel Act 1967, renationalising the industry and bringing fourteen big private companies – about 90% of UK production – together as the British Steel Corporation, with a workforce of 268,500. This creation would see enormous changes over the next fifty or sixty years and, one way and another, I would watch them take shape at quite close quarters for a considerable amount of that time.

Prime Minister Wilson announced the UK's application for membership of the European Economic community (EEC). Britain and France jointly made supersonic aircraft for the commercial market. Concorde was launched at Toulouse in 1967, but did not land at Filton, Bristol, for another two years. Maybe that was as well because of two serious air crashes. In the first, seventy-two passengers perished on a British Midland flight which crashed near Stockport when attempting to land at Manchester on a June Sunday morning. The second was when a Toulouse-built Caravelle on a scheduled flight from Malaga to London flew into Blackdown Hill on approach to Heathrow. Thirty-seven on board died with a suggestion that the design of the altimeter may have contributed because an indication of 6,000ft may have been read as 16,000ft.

A navigation error lay at the heart of the grounding of the 120,000-tonne capacity Torrey Canon between Land's End and the Scilly Isles when the BP-chartered vessel was making its way to Milford Haven from Kuwait. About 50 miles of French and 120 miles of Cornish coast were contaminated. Around 15,000 sea birds were killed, along with huge numbers of marine organisms, before the 270 square miles slick dispersed. Much damage was caused by the heavy use of so-called detergents to break up the slick.

The Ford Escort replaced the Anglia and Barclays Bank installed the UK's first cash machine at its Enfield branch.

St Pancras station was Grade I listed (see photo on page 140).

Briton Ferry Athletic soccer team

Loughborough Colleges postgraduate soccer team

1968

I had considered doing a Masters' Course at Birmingham on *Government and Industry* and was awarded a place. My shaky start to the final Politics exams resulted in my tutor telling me that if I had started them as well as I finished then.... Instead, I decided to apply for a postgraduate teaching course in education and physical education at Loughborough Colleges.

My start at Loughborough in the autumn was eventful for the wrong reasons. In those days it was customary to pack your belongings into a large trunk and send them to your accommodation address by a carrier of some kind. My belongings included all my athletic gear, essential items for the practical sessions one did every day as a physical education student. I lived at The Lodge in the small village of Lockington with four or five other non-PE Loughborough College students. It took three weeks for my distinctive African violet coloured kit to arrive. Nissan's workwear colour turned out to be a very similar colour some thirty years later.

Each day I would ask Mrs Dale, our landlady, 'has it arrived yet?'. The explanation for the delay was all too evident when the trunk eventually arrived: it was addressed to 'The Lodge, Lockington, Leicestershire' when the correct address was Derbyshire! Postcodes had already arrived in Derby in 1967.

The postgraduate syllabus comprised a theoretical and practical study of PE and an academic subject. The timetable started off with a 2-hour swimming session on Monday morning. Part of the course entailed survival and lifesaving tests. I passed the latter but, ironically, failed my personal survival because I just could not inflate my pyjamas to float.

1969

Teaching practice was at Bemrose Grammar School in Derby. It lasted for about six weeks from February 1969. By this time, I had acquired my first car, a Ford Anglia 100E. (For car enthusiasts, I need say no more!). It not only took me to Bemrose; it also ferried some of my fellow students living in the Lodge to the schools in which they were placed. In those days many motorways were incomplete, so my trip to Loughborough was via a poor heads-of-the-valleys road and up the A38 right through Birmingham. The Rock at Brynmawr was a particular challenge, especially during rainfall, when the air-operated wiper blades might slow down to a complete halt.

Loughborough was tough for a postgraduate PE student because of the combination of lesson preparation, physical activity and travelling, for both PE and another subject, was so demanding. At the end of the academic year, I was so exhausted I missed out on the sailing module at Grafham Water.

Career decisions were needed, but I did not make them because I was hesitant about teaching: I thought that teaching PE professionally would spoil my enjoyment of amateur sport. Lynn Davies, the Olympic long jump gold medallist at Tokyo in 1964 had started to train at the University track. When I left Neath Harriers to join Swansea, I met him there several times and trained together.

Again, I took several summer jobs – coaching athletics at the Afan Lido, labouring on a building site, process work in a bakery and, in September, at the steelworks in Port Talbot. When I worked in the steelworks during the summer holidays, it was not just about being a student and getting pocket money. I really enjoyed it all and people can't understand why I should have taken the path I did. Maybe it is genetic and social! Both my grandfathers were mechanical engineers in industry, and my unconscious preference was to do something more practical than becoming, say, a lawyer or PE teacher.

The work this time was in the Cold Mill. The reason I loved working in Port Talbot was that people were good to me, almost without exception. Particularly so were my colleagues at the acid recovery plant in the Cold Mill where I started work on October 8th. Brian Williams of Aberavon was the lead operator on our shift. It was he, and Manager Dennis Jones, who encouraged me to make myself known to the Personnel Development Manager. This led to me joining the 1970 personnel graduate trainee cadre.

Charging scrap into a basic oxygen (BOS) converter
(Amgueddfa Cymru)

Charging iron into a basic oxygen (BOS) converter
(Amgueddfa Cymru)

1970

I had already arranged with Camp America to go to the USA when I joined the 1970 BSC personnel graduate trainee cadre in May. The Personnel Development Manager would honour the arrangement, providing I agreed to spend time there in studying how US Steel managed its personnel policies and procedures, with arrangements made with US Steel from the UK side.

One of my first activities as a graduate trainee was on the 12th May to see the Queen, accompanied by the Duke of Edinburgh, officially opening the new Tidal Harbour at Port Talbot capable of accepting vessels of up to 170,000 tonnes of dry cargo. At the steelworks she unveiled a plaque to mark the official opening of the new Basic Oxygen Steelmaking Plant.

Many students participated in the Camp America scheme because BUNAC, (the British Universities' North America Club) arranged paid work at children's summer camps with cheap flights to New York. Mine was via Shannon at just 99 dollars. I was offered Camp Kenico, near Danbury, Connecticut. There, I could deploy my PE

Port Talbot harbour (ABP)

teaching skills as one of the camp counsellors. In some respect it was Loughborough all over again, being a six-and-a-half-day week from dawn till dusk. The seventh was a half-day during which the Camp Director had offered us the use of the camp bus to go into nearby Kent village. He never kept his promise. Instead, several of my fellow counsellors decided to take the law into our hands: if we could not go to Kent, then Kent could come to the Camp – in bottles. As this was frowned upon, and the Director saw me as an instigator, he decided to put me on alternative duties.

The first new task was to escort an unwell child from camp to home in Bayside, in Queens Borough of New York, via the Tapan Zee Bridge driving the Director's huge, black, Chrysler 300. He insisted that I should take a route over the Hudson River between Tarrytown and Nyack in New York state. It should be mentioned that the kids were from rich families and counsellors considered some of them to be quite spoilt, but the child involved was ill, not spoilt. In effect, this task was not a final warning, but a silent notice of dismissal for we three conspirators. The next bus we got was the one to a nearby town called Parksville.

There, we found our way to another camp at Elko Lakes, to take up work for a catering company called *Servomation* to serve the camp's population. It was run by Trinity Episcopal Church of Wall Street. Through direct outreach work, advocacy, speaking in the pulpit and in public, and through philanthropy, the Church aimed 'to transform its neighbourhood by changing systems, structures, and attitudes for racial justice and lasting change'. The poor kids who benefitted at Elko were mainly black and their unspoilt behaviour was a strong contrast with the children at Camp Kenico.

Following this, it was time to get to Chicago and Minneapolis/Minnetonka by Greyhound Bus. Chicago was a stopping point en route to the Silloways home in Minneapolis. Tessie Silloway was a Briton Ferry born GI bride who had emigrated after World War Two. One of her two sons, Bobby, had offered to take me around

Trinity Church, Wall Street, New York

Minneapolis and Silver Bay *(Google Maps)*

Minnesota, from Minneapolis to Duluth via St Paul. Duluth is the most inland port in the world, and it lies on the largest in area of freshwater lakes, Lake Superior. He thought nothing about doing the 350-mile round trip in a day. Years later, in 1982, he would return to see Wales with his wife and children. That's quite another story.

I knew little about the place and the company we were to visit, Reserve Mining Corporation's taconite mine at

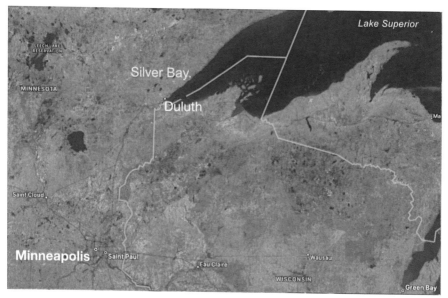

Silver Bay, near Duluth. I knew even less about its environmental problems. The taconite mined there is a low-grade iron ore, which can be crushed, concentrated, and pelletised for feeding blast and other furnaces. Silver Bay city itself (3,500 residents) was created during the mine's development.

The beginnings of one of the classic cases of environmental conflict was well underway at the time of our visit. In 1947 the company received permits to dump up to 67,000 tons per day from its operations into western Lake Superior. After decades of investigation and lawsuits, the corporation stopped the dumping in spring 1980.

By the late 1960s, major signs emerged that the lake, and potentially local residents, were being harmed by the waste discharges. The first permits were given on the condition that it would strictly comply with all provisions to ensure the discharges posed no harm to the lake or surrounding inhabitants, but later Reserve was allowed to increase its use of Lake Superior water, thus increasing its dumping of waste taconite tailings.

The first sign was the obvious discolouration of the water. Tonnes of tiny fibres resembling asbestos, a carcinogen, were dumped into the lake each day, turning the water into a green slime. Second, several local citizens made the link between the company's discharge and possible harmful effects on humans when they read about a discovered connection between cancer and the asbestos used to polish rice in Japan in the 1960s.

By June 1973, after completing several studies at the request of these local citizens, the U.S. Environmental Protection Agency (EPA) concluded that the drinking water of Duluth and other communities nearby was contaminated with asbestos-like fibres, which might cause cancer.

The Reserve Mining Company claimed that it was impossible to dump the tailings on land. During the trial, however, subpoenaed documents showed that Reserve had already examined dumping the sediment on land as an alternative method of disposal.

Both the Federal Government and Reserve relied on a prestigious lineup of scientists to buttress their case. After months of testimony, Judge Lord decided that dumping the tailings into Lake Superior posed serious health and environmental threats. In April 1974 he ordered the plant shut down.

> *I am not aware of a constitutional principle that allows either private or public enterprises to despoil any part of the domain that belongs to all of the people. Our guiding principle should be Mr Justice Holmes' dictum that our waterways, great and small, are treasures not garbage dumps or cesspools.*

The plant was closed temporarily, but a Federal Appeal court allowed Reserve to reopen the mine and to continue dumping in the lake until the company could find a new disposal method. In 1980, Reserve began to deposit waste in an inland pond, a practice that continues with the companies that mine taconite today.

The ruling in *United States of America v. Reserve Mining Company* was considered a landmark decision. It gave the E P A broader powers to regulate corporate pollution, a practice was unheard of before the lawsuit.

With regards to environmental protection, at this point it is worth mentioning another firm from Minnetonka: Cargill, which has many agriculture-related operations in the Marches, particularly Herefordshire. As a private company, it is not required to release the same amount of information as a publicly traded company and, as a business practice, keeps a relatively low profile. In 2019 Congressman Henry A. Waxman called Cargill 'the worst company in the world' and said it drives 'the most important problems facing our world' (deforestation, pollution, climate change, exploitation) at a scale that dwarfs their closest competitors. Today, Cargill is allegedly buying soya, cocoa, corn and more from farms linked to the destruction of the Amazon rainforest, slave labour, and stolen indigenous land. Today, Cargill is using wind assisted bulk carriers to help decarbonise its ocean-going shipping fleet.

Time went on and nothing had happened from Port Talbot about my visit to US Steel, so I decided to take matters into my own hands by dealing directly with US

35

Steel in Pittsburgh. I was asked to report to the company's Director of Policy and Programme Planning at their Pittsburgh HQ at 600 Grant Street. There, I spent a few days with a reasonably free rein to examine documents and ask questions to find out how an organisation like US Steel managed its personnel policies and procedures. It was exciting to see the Pittsburgh Pirates professional baseball team play St Louis Cardinals at Three Rivers Stadium in early September.

Although Gary Works and Burns Harbour were among the newer lakeside steel plants to the east of Chicago on Lake Michigan, I instead went on the Penn Central train to Fairless Hills of US Steel in the far east of Pennsylvania and Lackawanna of Bethlehem Steel on Lake Erie in New York at the end of September.

Fairless Hills township began in 1951 with the development of prefabricated homes built by a wholly owned subsidiary of US Steel with a loan of $50 million. Fairless Works, on the Delaware River, employed most of Fairless Hills' homeowners. The 1,600-acre site was the largest steel mill ever built all at once. Such prefabricated homes once helped to address the UK's post-war housing shortage. They could well do likewise today.

US Steel's Fairless Works during construction

Iron ore cargo arriving at US Steel's Fairless Works

Bethlehem Steel's Lackawanna plant was another that was part of the demise of the American steel industry. Located on the shores of Lake Erie near Buffalo, its Lackawanna Plant was once the fourth largest steel mill in the world. In 1922, the second largest steel company in the United States, Bethlehem Steel, purchased it and spent over $40 million in repairs and updates to the plant. It remained a large competitor in the steelmaking industry as a leading manufacturer of rails and sheet piling.

At the start of World War II, Bethlehem focused its operations on steel plate production for ships and tanks, as well as structural steel for the military. During that war it became the world's largest steelmaking operation employing over 20,000 workers on its 1300-acre site. Its prosperity lasted into the 1970s, but in 1977 the plant began to make cuts in its production and workforce due to decreased demand for steel and the sudden onset of steel imports from foreign markets. The company also disliked the high tax rates of the state of New York, and did not want to spend the millions of dollars in air and water pollution abatement which were required. The company instead built a new facility in Burns Harbor, Indiana, and stopped investing in new steel production methods at Lackawanna. Although the plant was profitable most years from 1970 to 1981, the company closed most of it by 1983 due to rising operational costs and the decreased demand for steel.

Prime minister Wilson was defeated in the June 1970 General Election by Edward Heath.

Prime minister Harold Wilson was defeated in the June 1970 General Election by Edward Heath

1971

In Place of Strife was a UK Government white paper which appeared in 1971. It proposed an act to use the law to reduce the power of trade unions in the United Kingdom. Proposed by the Secretary of State for Employment and Productivity, Barbara Castle, it was never passed into law. Amongst its numerous proposals were plans to force unions to call a ballot before a strike was held and establishment of an Industrial Board to enforce settlements in industrial disputes.

It was influential in the Trade Union and Labour Relations (Consolidation) Act 1992, (TULRA) including the requirement that strike action could only take place after a trade union ballot would later become a key component of TULRA.

Successive governments had already become involved in legislating and regulating terms and conditions of employment. In 1963, under the Conservative government of Harold Macmillan, the Contracts of Employment Act was passed, requiring employers to give workers a minimum period of notice when terminating their contracts and to give written particulars of any verbal contract when a written contract was not provided. One of the first pieces of legislation enacted by Harold Wilson's 1964–70 Labour government was the Redundancy Payments Act 1965, requiring employers to consult unions at the workplace in advance of decisions to terminate workers' contracts on account of redundancy. Employees were also given a statutory right to both notice of redundancy and substantial financial compensation.

The Labour government also fulfilled the long-standing ambition of the trade union movement by enacting the Equal Pay Act 1970, requiring employers to pay women the same wage as men for the same work. Though more narrowly drawn than many trade unionists had lobbied for, the Act nevertheless established an important principle. It was supplemented and broadened through collective bargaining and strike action in which women proved notably determined and were well supported by their male colleagues and union officials, e.g. the strikes of women sewing machinists at Ford's Dagenham and of workers at Trico's windscreen wiper factory in West London. This Act, together with the Race Relations Act 1968 and the Sex Discrimination Act 1975, provided the foundation of positive rights for women (and men) workers of all ethnic origins to be treated equally at work, both by employers and their fellow workers.

The General Council of the TUC took part in the National Economic Development Council (NEDC) but firmly declined, however, to co-operate with Macmillan's other initiative, the National Incomes Commission (NIC). Labour's attempt to launch a prices and incomes policy in 1964 with the establishment of the National Board for Prices and Incomes (1965–1971) proved to be another failure along with failures by the succeeding three governments, the Conservative Heath government of 1970–74, the minority Wilson administration of February–October 1974, followed by the Labour government of 1974–79.

1971 was a year of optimism in the UK steel industry because BSC's 1970–71 Development Plan and strategy envisaged increased production volumes, attributed to a rising demand for investment goods, consumer durables and cars.

By the time of nationalisation, the Steel Company of Wales (SCoW), like the rest of the British steel industry, was in financial decline. The company was beginning to address some of the fundamental issues that it faced such as building a deep-water harbour and converting steel-making to the BOS process. These reduced the cost of steelmaking and allowed the company to compete with the new strip mills. Only tentatively was it beginning to address its overmanning.

Port Talbot had a central personnel organisation

comprising industrial relations, staff development, planning and services functions to support the onsite personnel activities in each part of the works – coke and iron, steelmaking, hot and cold rolling, and engineering. Initial placement of graduate trainees was to the central organisation, in my case to education and training, a part of the staff development function. Postings elsewhere around personnel would follow, combined with a schedule of formal management training courses with other graduate trainees at the divisional training centre at Llangattock Park, near Abergavenny. Training runs along the nearby Brecon and Merthyr canal bank helped establish sufficient fitness to perform a long jump of 7m 14 for Swansea Harriers in 1971.

In 1971 the Abrasive Wheels Regulations were introduced, and one task in the education and training centre was to support Training Officer, Bill Bradley, to organise training in the safe use of abrasive wheels, for which he also prepared a technical booklet. The safe use of gases for welding and cutting skills was another example from the ongoing course schedule. Yet another was to organise both on and off-site courses for mobile plant operators of equipment like Caterpillar, Cummins and Allis Chalmers. To see what others were doing on operator training, Bill Bradley and I went to Dalziel Plate Mill in Motherwell to look at its operator training systems.

Swansea Harriers Athletics team, 1971

1972

In May 1972 the Minister for Industry, Tom Boardman, announced that steel industry investment would be concentrated on the five main plants without the creation of another on a greenfield site.

A decision of immense strategic importance taken was to import Llanwern's iron ore through Port Talbot Harbour. It required a third unloader at the ore jetty, extensions to the stockyard and rail loading equipment, with Llanwern meeting the rail transport costs. Importing the ore through Port Talbot gave Llanwern cost advantages over Ravenscraig. It also reduced the costs per ton of discharged material for Port Talbot. (In 1976 BSC signed a

No 4 Blast Furnace, Port Talbot

£19 million contract with British Rail to move up to 59 million tons of iron ore to Llanwern from Port Talbot over twelve years, involving up to seven trains daily, six days a week). The additional transport costs weakened Llanwern economically in relation to Port Talbot and made it more vulnerable overall.

The 1970s were witnessing the growth of the Japanese and Korean industry and growing exports of steel and automobiles were threatening domestic industries in both the USA and the UK. It would be another decade or more before the Japanese automotive industry would invest significantly in these countries to offset the loss of its own industries. In 2024 the coup de grâce occurred when Nippon Steel attempted to buy what was left of the US Steel Corporation.

Blastfurnaceman

The UK car industry in the 60s and 70s

The management of Great Britain's largest carmakers proved slow to adapt to changing markets and were sometimes stymied by their workers as well. Output at overmanned plants was hit by constant labour disputes from the 1950s, making them unproductive and unprofitable. British firms lacked the flexibility to compete abroad, even as European manufacturers began targeting the UK market with exports of right-hand drive models. British vehicles often failed to meet the technical standards required by export markets.

The UK government tried to come to the rescue, engineering a merger of the country's remaining large manufacturers in 1967, combining the Austin, Morris, Standard, and Triumph lines to create British Leyland as a national champion. Cars, unlike other manufacturing industries in decline such as steel or shipbuilding, represented more in terms of glamour and prestige. Failing motor firms not only dealt a blow to the economy but to the UK's ability to hold its head high. When government ministers intervened, as they often did, it was not just to save jobs but to rescue Britain's respect.

Upsizing did not solve the problems at British Leyland. After losing its market lead to Ford, and running up huge additional losses, it was nationalised in 1975. It was a lame duck, but insolvency was unthinkable to ministers: it was too big to fail and too strong a symbol of national status. Governments used the motor industry to implement economic policy, forcing firms to put factories in deprived regions and giving grants to create jobs.

   British Leyland was eventually sold to British Aerospace in 1988, resold to BMW six years later, handed to a UK consortium in 2000, and went broke after five years with Chinese firms buying its plant and designs. Even the prestige brands of Rolls-Royce and Bentley, nationalised in the 1970s after their aero-engine parent went bust, were bought by BMW and Volkswagen, respectively. Jaguar and Land Rover were acquired by Ford and then sold to Tata of India, while Peugeot closed the plants it had purchased from Chrysler. Bentley was bought by Germany's Volkswagen in 1998.

Nowadays, the UK has plenty of car manufacturing. In the 1980s, Nissan, Honda, and Toyota built plants in Britain, importing Japanese production techniques instead of Japanese-built vehicles. BMW constructed a new UK plant to make a relaunched Mini and multi-branded Stellantis still builds cars in Britain under the Vauxhall logo.
　While 80 per cent of cars purchased in the UK today are imported, at least three-quarters of cars produced in the UK are sold abroad.

1973

My attachment to staff relations did not work out so well. A sharp rise in energy prices had marked the beginning of a decade of inflation that rarely fell below 10% because of a tenfold increase in oil prices. As part of a wage-price spiral, the staff unions had negotiated a complex pay settlement to reflect the situation. Individual changes to pay were done manually and I was tasked with doing a batch of them. After only a cursory demonstration and explanation of what was required and how to do it, I was set to start. There was a high risk of errors carrying out the process that had been outlined to me, and I made them. Eventually, a senior colleague showed me a simple template that would reduce the risk of error and whose availability I should have been informed of before starting.

In February 1973 the Government published a White Paper *British Steel Corporation: Ten Year Development Strategy*. BSC recognised that it had to minimise costs and meet growing quality requirements. To do so, it was essential to use richer foreign ore transported in large bulk carriers to a deep-water ore terminal, use larger capacity blast furnaces, and BOS steel production to reach capacities of up to 6 million tonnes/annum. Accordingly, during 1973–74, Number 5 and Number 3 blast furnaces were relined and enlarged so that the former could produce 23,000 tonnes/week and the latter 12,000 tonnes/week. Number 1 furnace was also relined and began its fifth, and last, campaign in August 1975.

Despite my difficulties in Staff Relations, or maybe because of them, my first permanent appointment came as Training Officer – within the personnel section for the Coke and Iron works under Ken Wellington. Previously, in 1969, the National Union of Blast Furnacemen (NUB) at Port Talbot had gone on strike about their pay and conditions and some repercussions persisted. One was a concern about a lack of understanding by less-experienced staff of the processes where they worked and the risks to which were consequently being exposed.

An early job, therefore, was to work on a preparing standard operating and safety procedures for blast furnace stoves after the reline of Number 5 furnace. This was done based on the part time secondment of Gerald Williams, then an assistant blast furnace manager. Approval of the work was sought by both the NUB and the engineering union (AUEW) convenor, Hop Jones. One reason for this degree of attention is that blast furnaces produce carbon monoxide (CO) as a by-product. It is harmful when breathed because it displaces oxygen in the blood and deprives the heart, brain and other vital organs of oxygen. It is heavier than air, colourless, odourless and tasteless. Large amounts of CO can overcome you in minutes without warning, causing you to lose consciousness and suffocate.

The furnaces had a rescue team should any incidents occur, especially involving the gas. A recent incident had indicated that the team required refresher training. Laurence Houten, the company's Fire Officer, recommended that suitable training could be done at Lancashire Fire Brigade's Training Centre at Chorley, where smoke chambers were available. Organising this in liaison with Chorley became another important role for me.

Coke and Ironworks rescue team badge

To develop background knowledge in support of on-job training, Joe Rodgers, Sinter Plant Manager, offered to deliver a City and Guilds Advanced Operators Certificate in Ore Preparation and Ironmaking. This took the form of on-site Tuesday evening classes during Autumn and Winter in Port Talbot Works Education and Training Centre in conjunction with Margam Technical College. It was advanced, too, but Bill Harrison, Works Manager for coke and iron, nevertheless, presented five successful candidates, mainly shift workers, with their awards the following June. One recipient was 18-year-old

City and Guilds
Ironworks course
presentation

Kevin Downey whose involvement in two blast furnace incidents in 2001 and 2006 would end so tragically.

Yet another round of visits took place, as part of a BSC corporate working group to research operator training. Despite good practices, this time to Corby and Scotland with Ken Wellington. Clydesdale and Clydebridge would close within five years losing some 2,400 staff.

The mortal coil:
Steelworker
sculpture

1974

On 1st June, twenty-eight were killed and thirty-six seriously injured in an explosion at Flixborough chemical plant. It led to a public outcry over process safety and a more systematic approach to safety in UK process industries. The Health and Safety at Work Act, which was already planned, came into force on 1st October to change the way all workplaces dealt with the management of both health and safety.

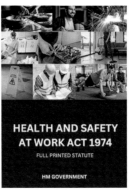

The Health and Safety at Work Act

The first training visit for the Blast Furnace Rescue Team to Lancashire Fire Brigade's Training Centre at Chorley took place in January. It was none too soon. Within weeks a death was sustained on one of the furnaces and the team were called for. Shortly after the Works Director invited the team to meet him, a further rescue successfully took place in the coke ovens when a bricklayer fell ill inside a cold oven.

In 1974 Lord Beswick, a Minister of State at the Department of Trade and Industry reported on the life expectancy of parts of the UK's steel industry. It was clear that it would be short for East Moors (Cardiff), Ebbw Vale, Shotton, Shelton and Hartlepool.

In Port Talbot things were more optimistic: training of the operators at the wagon loading station for the iron ore transshipment from the new harbour terminal to

Llanwern was in preparation. Trains, 27 wagons long, and sometimes triple headed, were loaded with over 2,000 tonnes of dense ore from the loading station bunkers for the 50-mile delivery. However, transhipment did not start until 1975 because of union concerns over manning levels at Llanwern.

Nevertheless, the reline and enlargement of blast furnace 5 was completed successfully.

There was talk of the Education and Training Manager's job becoming available and even my line manager encouraged me to consider applying for it. The vacancy did not transpire, but I was then considering a new challenge, and perhaps for the wrong reasons. Initially the challenge with the Iron and Steel Training Board (ISITB), was not a good choice, but within a short time the circumstances involved turned in my favour.

1975

My next fifteen years were spent at Manchester, the world's first industrial city. The circumstances which took me there were quite unintended: I was assigned to Manchester when I had been promised a post at Wolverhampton by my new employer, the Iron and Steel Training Board (ISITB). Its north-west office was in sleepy suburban Sale, whose closest steelworks was BSC's wheel forge in nearby Trafford Park, formerly known as Taylor Brothers.

After leaving Port Talbot works, between mid-1975 and 1977, £13.6 million was spent on improving the raw materials handling system to support the new sinter plant and enlarged blast furnaces. These improvements allowed an output of approximately 2 million tons of steel during 1978-79, using three rather than four blast furnaces.

Ironically, I spent my first few weeks' induction with ISITB back in Wales at the then, GKN's, Tremorfa plant in Cardiff. Work with the board was partly administrative, entailing the processing of statutory levy/grant applications, and partly practical in the form of doing checks on training programmes in the workplace. The admin was not particularly interesting, but the factory visits were, including to a firm called Manchester Steel (MS). The problem was that there weren't enough levy/grant applications visits to keep my interest alive in the northwest. Steelmaking at Irlam and Shotton had gone; Brymbo Steel's demise was in prospect, leaving just some finishing and smaller private concerns. An opportunity to deliver some residential supervisory courses on safety in blast furnace practice and continuous casting at Sandsend, near Whitby, was most welcome and rewarding for me because it sustained my interest.

Co-Steel of Canada had opened an early 'mini mill' at Sheerness in Kent in 1972 which could produce 188,000 tonnes a year. The plant was to have a chequered history because of ownership complications over the next three decades. Such mills also appeared in south Wales: GKN commissioned two 90-tonne arc furnaces at Tremorfa, Cardiff in 1976; Alphasteel built two 100-tonne furnaces in Newport and Duport Group commissioned two 125-tonne furnaces in Llanelli in April 1978–79. The latter made 10,000 tonnes a week on fifteen shifts.

EAF has tapped steel into ladle ready to cast into billets (Manchester Steel)

An electric arc furnace operator at Rotherham (Sky News)

1976

Another mill was nearing completion at Bradford in Manchester. It was being built for Manchester Steel, which was part of Elkem A/S, a Norwegian Company who had other metals interests including two aluminium smelters and ferro-alloys operations in Norway.

Two miles north-east of the city centre, Bradford, and nearby Beswick, was an area of heavy industry and closely packed terraced houses. Bradford lay to the north of Ashton New Road and to the south of the Ashton canal. What is today called Alan Turing Way to the west, was then known as Grey Mare Lane. Johnson and Nephew's steel and copper wire works operated there. Close by were Bradford pit (closed 1968) which supplied coal to Stuart Street power station by underground conveyor. Nearby areas such as Clayton, Miles Platting, Gorton, Openshaw and Newton Heath supported mainly chemical and engineering, rather than textile, activities.

The plant comprised a 60-tonne Demag electric arc furnace and four-strand continuous billet casting. Such mills were those that melted up to a million tonnes of scrap per year. Products from the mills ranged from billets to strip, bar, rod and light sections. Mini mills were regarded as a more flexible steelmaking process than the blast furnace route and were in many ways considered to have a more promising future. Operationally, their first trump card was that they used the more efficient continuous casting process which some plants such as Port Talbot and Llanwern had yet to introduce. The second trump card was that electric furnaces could be more easily shut down than blast furnaces.

1977

The mini steel plant was already in operation when Manchester Steel, already regarded as progressive locally, invited me to join them from the Training Board. I jumped at the opportunity. The offer came after I had spoken for the ISITB at a Manchester Steel joint management-union weekend seminar at Wetherby. Such seminars aimed to integrate staff who had come from Irlam Steelworks and the adjacent rod and wire mill, which had recently been purchased from Johnson and Nephew (J & N) The latter was a well-known and long-standing British Company, but one which needed modernising. It had constructed the world's first continuous rod rolling mill in 1862 and became one of the earliest suppliers of wire to the submarine cable industry.

The presentations at Manchester Steel's two staff Conferences at Wetherby were done whilst I was still an ISITB employee. My subject was the value of proper training. To demonstrate this, I used the example of how to disassemble a constant velocity joint. The aim was to show that 'it is easy when you know how'. All attendees, including the Managing and Finance Directors, were asked to disassemble the joint and all failed. As tutor, I simply explained and showed what was required to do so and everyone then succeeded with ease. I have little doubt that it was that presentation that resulted in an offer of a job at Manchester Steel.

Today I was reminded of the importance of 'know-how' when my task was to remove the hard drive from a *Dell Inspiron* laptop. All it needed was the removal of two screws and the drive would easily slide out of the machine. The tutor that provided the 'know-how' was *YouTube*, which told me that it was only necessary to remove two screws. Had I not known that I was liable to have removed over twenty, and perhaps still failed in the task. This was a great example of 'How to Design Everyday Things', as advocated by cognitive scientist Professor Don Norman. Years later, the example of error prevention (e.g. removing screws unnecessarily) but instead, during the misfuelling of light vehicles, another everyday task, would challenge me yet again.

Since the coming to force of the 1974 Health and Safety at Work Act, the company made great efforts to make the Safety Representatives and Safety Committees Regulations a success. A joint company-union Safety Charter was introduced and much work on standard operating and safety procedures followed from this. It worked well, and the new company was enhancing its

Manchester Steel's Safety Charter

positive reputation. There were several examples of this. The purchase of Johnson and Nephew (J & N) offered an early opportunity for management and representatives of both MS and former J & N to visit Oslo and meet our counterparts. Everyone was expected to work to the company's *Principles of Organisation and Leadership* which stressed accountability and responsibility.

Meanwhile, despite an occasional setback, such as a major furnace transformer failure, productivity increased strongly, especially in the rod rolling mill when Paul Monk was introduced as Manager. He increased people's competence and confidence which enabled new practices to be introduced, and to drive up production volumes and quality standards.

The Bradford area of Manchester was, of course, highly industrialised and, in 1960, was claimed to have been 'the most atmospherically polluted area in the world'. In that respect it was a more recent version of the lower Swansea valley. It was unsurprising that one of my first tasks was to co- operate with the Health and Safety Executive's Alkali Inspectorate to make an onsite film one weekend called *The Sky is a Canvas*. It was mainly set in the Steel plant and Rod Mill control rooms to show that, with efficient dust and fume extraction, it was possible to run a noise and dust-free steelworks close to a city centre. In this connection, the company had already established a good relationship with Salford University Acoustical and Engineering Department and benefitted from its expertise, and positive public relations.

By the end of the year, I was assigned to deliver Elkem's *Organisation and Leadership* courses. Their purpose was for supervisors, managers, and senior representatives to understand Elkem's management principles, particularly regarding the importance of, and the difference between, accountability and responsibility. Although I had attended some Norwegian language classes I delivered the *Organisasjonen og Ledelse* seminars in English!

1978

When I visited my parents at Christmas time, my father's health was evidently in a serious condition, suffering from stomach cancer. His death in February was, nevertheless, a great shock. I was at work in the rolling mill when a phone call came, and it was Paul Monk, the Rod Mill Manager who told me that there was an important call and then a journey to make.

By this time the company had created an effective structure for practical safety and training using the experience and knowledge of long-serving supervisors Bill Prescot, on Rod Mill matters, and Stan Barrowcliffe on steel melting and continuous casting. Bill had been a shift production supervisor at Johnson and Nephew and Stan in both Sheffield and at Manchester Steel.

That journey was to my hometown of Briton Ferry. There, at Duport's Albion steelworks, alongside Brunel's dock, its open-hearth furnaces would tap for the last time and the mill would roll its last billets after eighty-five years during both war and peace.

The year of 1978 was therefore one of very mixed fortune. The so called 'winter of discontent' from November 1978 to February 1979 was characterised by widespread strikes by private, and later public, sector trade unions demanding pay rises greater than the wage limits the Labour government had set to control inflation. The cause was in the increases in oil price, but the workers, as usual, were blamed, especially by the right-wing press. Some of these industrial disputes caused great public inconvenience, but it should be remembered that it was exacerbated by the coldest winter in sixteen years, in which severe storms isolated many remote areas of the country.

Bidston's billet caster
(Manchester Steel)

1979

In May, Parliament re-opened with Mrs Thatcher as Prime Minister. The first pledge on the Conservative manifesto was 'to restore the health of our economic and social life, by controlling inflation and striking a fair balance between the rights and duties of the trade unuon movement'.

The Thatcher government therefore came to power in 1979 with privatisation as a minor part of its manifesto, but it became a central part of its ideology as the 1980s progressed. Many industries and utilities that had been nationalised in the Attlee government of 1945–51 were made into private companies: steel, railways, airways, airports, and aerospace; and, of the utilities, gas, electricity, telecoms and water.

Early proponents of privatisation believed that creating markets would ensure services were efficiently produced and high quality, and the need for regulation would 'wither away' as competition took hold. In practice this has not happened. These industries remain highly regulated – in several, regulators set the prices companies can charge and the amount they are expected to invest.

A referendum on devolution for Wales was rejected by 79% of voters amid regional divisions and cultural concerns.

In August 1979 Manchester Steel Ltd bought Bidston Steel. Bidston (population 10,000) is now a ward of the Metropolitan Borough of Wirral, but before local government reorganisation on 1 April 1974, it was part of the County Borough of Birkenhead, within the geographical county of Cheshire. Formerly part of British

Reinforced Concrete, the Bidston plant included an 80-tonne electric arc furnace, billet caster and a rod and reinforcing bar rolling mill. At approximately the same time as the Bidston acquisition, Stewart Metals of Openshaw, Manchester, was purchased to secure reliable and local scrap supplies.

This, however, was also the time that Mrs Thatcher introduced Ian McGregor. How desperate they became to kill off the coal mines and steel mills, and even, indirectly, the privately-owned Manchester Steel. They called it 'industrial re-structuring'. Sir Keith Joseph, the Secretary of State for Trade and Industry, had appointed Ian McGregor Chair of British Steel in 1980, and in March 1983, Head of the National Coal Board, as a vanguard of privatisation.

Ian McGregor
(Wikipedia)

In November 1979 the south Wales strip mills were presented with four alternatives by BSC management. The one chosen was Slimline, which would maintain both Port Talbot and Llanwern with reduced steelmaking, rolling and workforce. Slimline was believed to be the most effective long-term way to reduce the cost base for strip mill products. For Port Talbot, this still meant the cancellation of a British Steel development plan and a threat of many job losses.

Frank Field became MP for Birkenhead and spoke up for Bidston Steel in the 1982 House of Commons debate on the steel industry. He died in 2024 whilst this book was being written.

1980

The 1980s was anything but a quiet decade. The early years were characterised by civil unrest in Britain. In 1981 the most serious riots were in Brixton, London, Toxteth in Liverpool, Handsworth in Birmingham, Chapeltown in Leeds, and Moss Side in Manchester. Industrial unrest accompanied or followed them. So, too, did several major industrial accidents and natural disasters in Britain and worldwide.

January 1980 began with the first national strike by steel workers for more than 50 years, over a bid for a 20% pay increase when the management had offered 5%, and the threat of plant closures. Margaret Thatcher's Conservative government, elected in 1979, set out to cut British Steel's losses at a time of overcapacity in the industry, rising energy prices and a deepening recession. A deal was eventually struck after nearly 14 weeks.

In May, the government appointed Ian MacGregor as British Steel chairman to drive through its savage rationalisation programme. By the end of the year, Consett, Corby and Shotton steelworks were closed, with the loss of more than 20,000 jobs. Total employment in the industry almost halved between 1979 and 1981, from 156,600 to 88,200. Thames TV monitored the of this in a programme about Port Talbot called *Steel: the hidden cost* which calculated the direct and indirect effects of this as a loss of some 17,000 jobs.

The stoppage gradually spread to the privatised steel works, including Manchester Steel's plants. The BSC plants reopened in return for an agreement on working practices and productivity deals. Later that summer, 17,000 of the 24,000 South Wales steel workers were put on short time. By the end of 1980, BSC had completed the closure of several outdated and loss-making plants and reduced its workforce to 130,000 – compared with a total of 268,500 employees at the time of nationalisation.

Since BSCs main problem area was its strip mills, Port Talbot was a prime target. It resulted in manning reductions, the end of restrictive practices, a switch to flexible working and outsourcing. The local restructuring developed into an ongoing process rather than being a one-off exercise. The required changes initially took local management by surprise and resulted in a somewhat rushed process which met the demands of the politicians, but also bought time to develop a more considered longer term commercial strategy.

One aspect of these events was a shift in production from the south Wales strip mills to Ravenscraig, but this proved to be only a short-term measure to retain market share, because only Ravenscraig had an operational continuous casting plant. Once manning levels had been reduced, and Port Talbot's continuous caster was commissioned, production moved back to south Wales. The 1980-81 Corporate Plan followed this period of upheaval and was intended to introduce a more commercial approach to production.

Not long after the strike ended in April, Manchester Steel's safety committee won the Sedgwick/Bland Payne safety award, and the prize was that members would

World Safety conference delegates at Hoogovens, Ijmuiden

attend the World Safety Congress in Amsterdam. It included an optional visit to Hoogovens Steel, the firm which would later join a privatised BSC in both Corus and Tata groupings.

I afterwards learned to my embarrassment, from a Sedgwick director, that the new Company Secretary had moved Manchester Steel to a new insurance broker, and it was not Sedgwick.

Stan Smith, who had joined Manchester Steel as its new Chief Engineer from BSC at Scunthorpe, was keen to get even more specific safety training done. By November, the Joint Health and Safety Committee was running a three-hour off-the-job training presentation for Manchester's steelmaking, rolling and maintenance personnel. Plant isolation methods, safe working procedures and permit to work systems were emphasised and set out in a handbook for everyone.

Safe working procedures manual

1981

After commissioning the steel plant, efficiency improvements were sought through the installation of such technology as water-cooled furnace panels and oxygen injection. After the strike, a group even participated in the August Manchester Marathon. That followed a spell in Withington Hospital receiving traction after a football injury.

Manchester Steel marathon runners – between the Stag and the Shakespeare on Philips Park Road

In conjunction with Openshaw Technical College, staff were encouraged to sit for the City and Guilds Operator's Certificate in Mini Mills Technology. Evening classes were held in the onsite conference room with good results awarded all round. The syllabus was drawn up by the College in consultation with Bill Prescot, on Rod Mill matters, and Stan Barrowcliffe on steel melting and continuous casting. The pair of former production supervisors also looked after the work instructors for on-the job crane operations loader driving, and oxy-fuel flame cutting. Supplementing these initiatives were placements for Salford University chemistry and business studies graduates, like Owen Leeds, David Conry and Tom Murphy who sharpened up the company noise control programme.

The Shakespeare pub on the corner of Grey Mare Lane and Philips Park Road

1982

Early in 1982, although I spent much time at Bidston, it was possible, in conjunction with Openshaw Technical College, to encourage staff to again sit for a City and Guilds Operator's Certificate in Mini Mills Technology, but this time it was for the advanced version of the award. Once more good results were awarded all round.

The Falklands Conflict, from April to June, was a short, undeclared war between Argentina and Britain over the sovereignty of the Falkland Islands, South Georgia and the South Sandwich Islands. The conflict lasted 74 days and cost over 900 lives. It was a difficult time for all concerned.

Remember Bob Silloway, who took me around Minnesota twelve years previously? This was the year he, his wife and children visited Wales for the first time. Naturally, I was pleased to see them one day, to take them to local activities of interest. After thinking the children had exhausted themselves at Afan Argoed country park, near Port Talbot. At the beach they simply ran at full speed, fully clothed into the waves. It was the first time they had seen the sea! The second memorable event of 1982 was the Bolton Marathon on 22 August. Perhaps it was just memorable, if not truly pleasurable.

On 1st December, the House of Commons debated the steel industry. I had no idea what exactly was going on behind the scenes about Manchester Steel, but I was uneasy about some suspected changes. Therefore, I decided to look around and obtained an interview for a post at Minet insurance brokers in December. My last diary entry for my time at Manchester Steel would be a visit to Bidston on 6 January 1983.

Employee safety Booklet: *Your Company Needs You*

1983

Leaving the steelworks to join Aon-Minet in February was something of a move into the unknown. As a risk assessor, operating mainly from Princess Street, Manchester, and sometimes Leman Street in London, the work had some similarities with previous work. Minet had a large portfolio of textile clients, such as Tootal and Carrington Viyella, along with other non-textile clients, such as Allied Steel and Wire. After joining Minet, I returned to Manchester Steel for a day or two to assist my replacement, at the request of my Minet manager who no doubt saw it as a potential introduction to broker the company's insurance accounts.

   I enjoyed learning about the textiles industry and its processes, from spinning, weaving and dyeing to the manufacture of garments and home furnishing fabrics. Things started well, passing my insurance exams, but remote learning became difficult. I was disappointed that I could not be assigned to clients such as Allied Steel and Wire where I could apply my steel knowledge so, eventually, I decided it was time to move on. Co-Steel, of Sheerness on the Isle of Sheppey, approached me (via their insurance broker) to join them. However, I did not like the Canadian-owned company's aggressive approach to staff relations, so I declined the offer. The Company later became non-union (1992).

   After the re-election of the Thatcher government in 1983, Ian MacGregor was appointed as the head of the National Coal Board. His immediate background was in British Steel, where he had been appointed chairman in 1980 to drive through the savage rationalisation programme which had resulted in so many cutbacks and closures. His impact would be felt again, the following year in the coal industry.

1984

The year of 1984 will be known for a long time as the year of the Miners' Strike, because the strike according to some, was a turning point in British history. The 1977 Ridley Plan had outlined how a future Conservative government could confront the trade unions after the 1974 Miners' strike had brought down the Conservative government, leading to a new Labour administration. The plan included targeting specific unions, stockpiling coal in power stations, undermining union finances, and special police training to deal with picketing. Margaret Thatcher, elected in 1979, was determined to limit the trade unions which she saw as having too much political power. Coupled with her drive to privatise nationalised industries in the name of economic efficiency, the mineworkers became an obvious target.

In February 1984, Nissan and the UK government signed an agreement to build a car plant in the UK. Candidates for the plant were south Wales, south Humberside, Cleveland, the west Midlands, and Wearside. The following month a 799-acre greenfield site at Sunderland at Wearside, was chosen.

Soon afterwards, also in March, the National Coal Board (NCB) unexpectedly announced that the closure of Cortonwood Colliery in South Yorkshire would take place at the beginning of April that year, the Cortonwood miners walked out on strike, initiating one of the most significant industrial disputes of the 20th century. The strike rapidly spread through the coal fields, polarised the country, and left a legacy in the former mining communities and elsewhere.

At that time, I was in west Wales for Minet visiting several of Tootal's clothing factories in Swansea which produced mainly contract apparel under the brand of Slimma. Tootal was a client of Minet. A further visit to west Wales in June took me to several more west Wales factories. Just for the record, these were almost all outside of the south Wales coalfield in Llandovery, Lampeter, Newcastle Emlyn, Cardigan, Fishguard, Fforestfach, Llansamlet and Ystalyfera.

It was on the return trip to Manchester that the seriousness of the unfolding events hit home with me. I had stopped on a bridge on the Heads of the Valleys Road. From there I could see a long convoy of military-type vehicles heading north. What on earth was going on? Later I deduced that they were heading for Orgreave coking plant in South Yorkshire. I then wondered whether what was to happen in Orgreave would eventually adversely affect the future of these remote west Wales clothing factories.

At Orgreave, on 18 June 1984, approximately 5,000 pickets faced 6,000 police. In the violence that followed, sometimes referred to as 'The Battle of Orgreave', over a hundred miners were injured and ninety-five were arrested and charged with riot or violent disorder. In 1991, thirty-nine of the convicted miners were compensated £425,000 by South Yorkshire Police, for assault by the police, wrongful arrest, and wrongful prosecution.

Some miners began returning to work from September, with many more returning by January as union pay ran out and some found it hard to pay for necessities. The strike officially ended on 3 March 1985. Despite the miners' efforts, the government succeeded.
    The defeat of the strike led very quickly to the closure of most coal mines, a general deindustrialisation of the economy, the rapid privatisation of nationalised industries, the shattering of organised labour, growing

unemployment, the hollowing-out of mining and other working-class communities, and a steady increase in social inequality in British society. It marked, in a word, the end of twentieth-century Britain and the ushering in of twenty-first century Britain, characterised by speculative capitalism after the Big Bang of 1986, the dismantling of workers' protections and the rise of the gig economy.

It was not just the year of the Miners' strike: if you lived in the Indian city of Bhopal, 1984 was also a hellish year because it was the site of the world's worst industrial disaster. A leak of toxic methyl isocyanate during a pesticide making process at Union Carbide's chemical plant exposed some 500,000 persons to the gas. 3,787 officially died but the true figure of both immediate and later deaths and permanently disabling injuries is still uncertain.

Striking miners

Miners strike poster
(Wikipedia)

1985

After the miners' strike was over, the government turned its attention to the steel industry.

Nothing of great interest was happening with Minet or its clients, although I attempted some more specialist insurance exams. I had not prepared for them properly, so I was out of my depth. The culture of the insurance industry was quite different from manufacturing industry. I was tiring of it when things were happening concerning Manchester Steel.

The House of Commons debate on its closure started at 11.40 pm on 5 June 1985. Manchester Steel Ltd was the holding company for Manchester Steel, Bidston Steel and Stewart Metals. The parliamentary debate is a case study of de-industrialisation, and of east Manchester in particular. It included accusations of government betrayal of a private, profit-making, organisation and indifference to both local and national interests and accusations against the owners of Manchester Steel Ltd of secrecy and duplicity.

I returned to Manchester Steel to see the last cast of steel billets before the closure of steelmaking during a night shift in December 1985. Everything went smoothly and professionally; although I no longer worked there, I was proud. For most of the staff, this had been their second chance, having been made redundant earlier from other workplaces. The sad situation comprised a complete waste of skill, knowledge, and commitment.

Today my former office, the converted Stag pub in Philips Park Road, Bradford, M11, lies in the northern goal area of Manchester City's Etihad Stadium. The surrounding areas have changed entirely under the name Eastlands – a name that those born and bred in Bradford

The Stag pub,
Philips Park Road,
Manchester
*(Pubs of
Manchester)*

and nearby find it hard to accept That is because their livelihoods were sustained by work at such places as Clayton Aniline and Bradford Pit rather than the National Squash Centre, Velodrome or Etihad Stadium that today occupy Manchester's M11 postcode.

Across the Pennines, in the city of Bradford, Yorkshire, fifty-six football fans perished in a fire at the Bradford City v Lincoln match. A stand that had been officially condemned and was due to be replaced with a steel structure after the season ended, caught fire during the match.

The last cast at Manchester Steel was a very sad occasion. I also felt a change was needed from Minet, and it took the form of a fly/train trip across Siberia. It was run by Intourist, created by the USSR in 1929 to attract tourists and promote the country. The flight from the UK from Manchester to Moscow was followed by flights

At Khabarovsk

within the USSR to Leningrad, Irkutsk, and Bratsk, both by plane and helicopter. It included a visit to Bratsk Hydroelectric generating scheme, capacity 4,500 megawatts. The train trip from Irkutsk to Khabarovsk took three days and nights. Having safely returned to Manchester Airport on 14 July, and a visit to Lloyds of London on 17 July, it took a month to realise how hazardous the Russian flights had been.

On 22 August, a British Airtours flight from Manchester to Tenerife suffered engine failure which generated a fire causing the loss of the aft exits. Eighty-two of the 131 passengers and six crew survived but Brian Taylor, Manchester Steel plant's mechanical engineer and his family did not. Most deaths were due to smoke inhalation. An aviation analyst said the accident was 'a defining moment in the history of civil aviation' because it brought about industry-wide changes to the seating layout near emergency exits, fire-resistant seat covers, floor lighting, fire-resistant wall and ceiling panels, more fire extinguishers and clearer evacuation rules.

The Trans-Siberian railway route *(Trans-Siberian Experience)*

In March 2021, Peter Finnegan, who was the last Company Secretary of Manchester Steel and had been a good friend before he left Manchester, wrote to tell me of the death of Ken Knaggs, who was the Managing Director of the company at closure. Here is part of my reply:

Dear Peter,

Recently I read the story of a Bevin Boy, conscripted from a Dorset farm to work in a mine in a Monmouthshire valley. I remember you sending me a copy of your ex-Bevin Boy father's payslip from Bradford Colliery. Has anyone recorded the industrial history of that part of east Manchester, of Manchester Steel, Bradford Pit, of Laurence Scot Electromotors, of the Finnegan family? The changes have been so profound and so fast that it is hard, today, to believe what it was like not-so-long ago.

The days of oil and coal are gone, replaced by wind and photo-voltaic, something you are all too well aware, I am sure. But the internet did not arrive soon enough to capture the days of Manchester and Bidston Steel, just an online Hansard entry about Manchester's demise and a sad and disappointed ex-employee's expression of what had been lost at Bidston. As far as I was concerned, what the Norwegians started at Manchester was continued by the Japanese at Sunderland. In between was Minet, which did not work for me, but moved me to Courtaulds: an interesting experience. That experience was to witness the death throes of another great British manufacturing enterprise. Just like ICI, first demerger, then dismemberment and finally death. Financial capitalism worked well for much of London and the home counties, but since the Thatcher era it has done little or nothing for the rest of this place. And now we have the self-destruction of Brexit which threatens both.

Then, later in 1985, an advert in the *Manchester Evening News* offered a possible opportunity to join Courtaulds Textiles, whose HQ was at Walkden in Manchester. I accepted, to start work in November, with the advantage of no home move being involved. The winter had an early start with the month, being the coldest in Central England since 1925. Courtaulds was a United Kingdom-based manufacturer of fabric, clothing, artificial fibres and chemicals. It was established in 1794 and became the world's leading man-made fibre production company. It was 'demerged' in 1990 into Courtaulds plc and Courtaulds Textiles Ltd. The latter had been brought together in April 1985 and employed 37,000 staff when I joined as Textiles Health and Safety Manager at its Walkden HQ later that year. Reporting to the Director of Management Services, my position meant lots of travel throughout Wales, England, Scotland and Ireland, north and south. In England alone, the company had large concentrations of factories in Lancashire, West Yorkshire, and the East Midlands. It was organised into product groupings – Spinning, Fabrics, Textile Finishing, Linens, Contract Clothing.

Internally, my work entailed building diplomatic and productive relationships with the textile groupings, individual factories and the Courtaulds plc (Group) Safety team, based at Coventry. Our working logistics meant that Management Services staff would pass each other 'like ships in the night' on daily visits to the factories. Much travel was by car and enormous mileages were covered. I thought that the logistics situation was a good reason to have some form of periodic departmental meeting, in addition to any random, haphazard day-to-day contact, to share experiences and keep updated on events and the work of others. But we did not do so.

Nevertheless, it was possible to create a system of Safety Panels, based on Courtaulds Spinning's prototype, which became the cornerstone for Textile Group's Health and Safety arrangements. Each panel was chaired by a production director supported by a safety adviser. Their role was to guide management in protecting people, property, and environment. Panels worked closely with our insurers and brokers.

1986

January was the coldest since 1963 with widespread snow cover. The heavy snow and sub-zero temperatures that affected most of Britain during February continued into March and led to some peculiar events which required attention. At Brands Group's warehouse in Belper, the severe frost caused sprinkler pipes to burst, causing much stock damage. I attended once or twice with Mike Sykes, Chief Engineer, who came up with the solution of trace heating to the sprinkler pipework. Such an innovation was essential: nearby Buxton's highest temperature throughout that month was just one degree Celsius.

In April, Number 4 reactor at the Chernobyl nuclear plant in Ukraine, (then in the USSR), sustained a steam explosion and fires which released part of the radioactive reactor core into the environment, with the deposition of radioactive materials in many parts of Europe. It was the world's worst nuclear accident and the result of a flawed reactor design that was operated with inadequately trained personnel.

Bob Northey was another fellow member of the Courtaulds Textiles central team. He had much production experience in weaving operations, but his group role was now Quality. He advocated works such as Tom Peters' *In Search of Excellence* as part of a move to the ISO 9001 quality management system certification he also advocated. What I learned from Bob helped me better understand Nissan's philosophy and systems that will be covered later.

So widespread were the company's operations that my diary that year recorded three trips to Ireland, north and south, two to Scotland and, in England, Cambridge, Cumbria, Durham, Lancashire, Leicester, Northumber-

Standfast, Lancaster, a printed fabric producer

land, Nottingham, Warwick and all parts of Yorkshire.

Significantly in Wearside, in July 1986, the body, paint and final assembly shops at Nissan were ready to roll the first *Bluebird* car off its new production line. A single-union agreement had been made with the AEEU a year previously and Bluebird production would continue until 1990.

The structural change to the financial market in London in October, when the Stock Exchange's rules were altered was to be significant for the country. Dubbed *Big Bang*, its changes saw many of the old City firms being taken over by large banks, both foreign and domestic, and would lead to further changes to the regulatory environment.

Its effects were dramatic, with London's place as a financial capital decisively strengthened, to the point where, in 2006, it was arguably the world's most important financial centre. The boom resulted in the relocation of institutions into new developments in the nearby Isle of Dogs area, particularly that of Canary Wharf.

There is a debate in the UK about how far it affected the Financial Crisis of 2007–2008.

Aerial view of Nissan, Sunderland *(NMUK)*

1987

Following the Bhopal and Chernobyl major accidents, the public began to understand that major 'accidents' were really the result of failures in the design of complex systems and human factors failures. The capsize of the *Herald of Free Enterprise* when leaving Zeebrugge in March 1987, killing 193 passengers and crew, no doubt sharpened that understanding.

The car and passenger ferry was built with no watertight compartments and the ship left harbour with her bow door open, so the sea immediately flooded the decks; within minutes, she was lying on her side in shallow water. The official inquiry recognised that the negligence of the boatswain in not closing the bow doors played a part but placed more accountability on his supervisors and a general culture of poor communication in the ferry operator, Townsend Thoresen.

Courtaulds Spinning's Maple Mill, Oldham

Do you remember 15 October 1987? On that date I was at an employer's liability meeting with GRE/AXA at the insurance company's office in Ipswich. It overlooked Ipswich Town FC's Portman Road football ground. At the end of the meeting, as I looked down to the stadium, for no obvious reason, a feeling came over me that it was time to return to Manchester-and quickly! Soon after leaving, the heavens opened, and the wind rose. The Meteorological office weather reports that day had failed to indicate a storm of such severity that was about to occur. It later reported winds gusting at up to 100mph. There was massive devastation across the country and eighteen people were killed. I was just quick enough not to have joined them. Vision through the car windscreen was difficult with a high risk of aquaplaning on the 230-mile journey and I was so relieved to reach home.

Another trip that had to be made quickly was to investigate a 'fatality' at McIlroy's department store in Godalming. The management were being hyper-cautious in calling me. It turned out that the lady concerned had died of natural causes but the conditions in the store were in no way adverse for customer safety and I reported that there were no preventive measures that could have been taken to prevent a recurrence. That reassured a caring store management.

In 1987, Bidston's arc furnace was put up for sale and, by 1989, it was dismantled and rebuilt in China as the Zhang Jia Gang municipal steelworks. Other mini mill operators fared no better: ASW were bankrupt despite its reverse takeover of Sheerness Steel in 1988. Sheerness, which had opened on its dockyard site in 1971, and would be twice closed and reopened, was eventually demolished in 2015. In this case, Liberty House, who occupied the former Alphasteel works on Newport docks re-installed Sheerness's arc furnace there. Its claim now is to produce green steel by using a larger proportion of the 80% of the UK's scrap that is presently exported.

Courtaulds Spinning Safety Award

1988

Privatisation of the industry by the Thatcher government created British Steel plc, a FTSE 100 company. In 1989–90 the company went on to make a pre-tax profit of £733m with its UK steel workforce of 55,000.

The year started with a visit to Scotland, during which I received an early phone call from Alastair Strang, who chaired Courtaulds Contract Apparel Safety Panel. He rang to say that Colin Kilbourn, the group's Safety and Security Adviser had been found dead quite unexpectedly. The concern was the large amount of money that he was carrying, for which there was no obvious explanation. I knew of no reason why he was carrying so much cash. I never found out, and it was not a question to raise at his funeral.

The safety panel system was working well and introduced some novel ideas to involve staff. One such was the SNIFF campaign in the contract clothing group. SNIFF was an acronym meaning *Stop needles in fingers forever*. Victoria Channing, who was Occupational Health adviser in Courtaulds London medical department heard of this and contacted me with a view to creating a London Safety Panel for the showrooms and offices there. That was how we met.

Meanwhile, between 1988 and 1990, Nissan was expanding with the addition of a Plastics and an Engine Assembly shop.

In the North Sea, on 6–7 July, the Piper Alpha oil platform, 120 miles north-east of Aberdeen, exploded and collapsed under the effect of sustained gas jet fires killing 165 of the men on board as well as two rescuers. The accident was the worst ever offshore oil and gas disaster in terms of lives lost. The inquiry attributed it to inadequate maintenance and safety procedures by the operator, Occidental, though no charges were brought.

1989

A dangerous occurrence tool pace one spring morning on the M1 near Chesterfield. It happened to me on my way to a safety panel meeting at Langley Mill. Whilst driving southwards in the centre lane, there was a sudden bang. The back of the Vauxhall SRi car collapsed onto the roadway and, as it contacted the ground the rear nearside wheel hurtled past and was never retrieved. I managed to carefully guide the slowing car to the hard verge on the left, being careful to avoid anything coming from behind…. Driving across the Pennines from Manchester to Nottingham two or three days a week throughout the year is not recommended for career longevity.

An audit visit to the Desseilles lace-making plants in northern France led to some significant changes at the old-fashioned Calais factory. The town had become famous for its lace making in the 19th Century, when English lace makers, famous for their lace-making loom inventions, smuggled one of the new Leivers looms into France and set up shop at Saint-Pierre, as Calais was then known. The Anglo-French collaboration was immensely successful. It completely transformed the French lace making industry which had hitherto been completely by hand. Although made by heavy machines today, the beautiful product is most delicate. Indeed, Desseilles Lace enjoys protected designation status as a 'Dentelle de Calais-Caudry', which is exclusively reserved for Leivers lace woven in Calais or Caudry.

There was a downside to the seventeen-tonne English-made Leivers machines each of which have 40,000 moving parts (spelt Leavers in France to aid pronunciation). It was that, when they were housed in groups to optimise power, the lofts of the buildings in which they were

Lace factory of Calais-Caudry

installed were very noisy and the floors became oil-soaked after many years of use.

The audit report highlighted the levels of noise and accumulation of oil as unacceptable. Although it did not recommend closing the factory, that is precisely what was done. Desseilles then improved both the working environment and conditions and refurbished the wonderful machines in a completely new building.

The public began to understand, too, that some of the major 'accidents', like the Exxon Valdez incident in late March, were a strong reminder that we had a planet to protect. Spilling 37,000 tonnes of crude oil in one of the largest environmental disasters in US history, in an ecologically sensitive location, was a wake-up call for all.

The Hillsborough football disaster followed a month later. The major incident in which 97 fans died as the result of a crowd crush had no such environmental implications but it became an important wake up call for the administration of justice in Britain.

In October, Courtaulds Textiles held its first environment conference at Kenilworth. Twenty days later conference Chair Martin Taylor became Chair of newly-demerged Courtaulds Textiles.

1990

At its outset, in 1985, Courtaulds Textiles made £60 million profit and the plan was to expand overseas. That never really happened to the extent desired. Instead, unlike the more sudden closure of Manchester Steel, the demise of Courtaulds Textiles was to be death by a thousand closures. Courtaulds plc did not fare much better after its demerger.

By 1990, things were starting to go wrong for me, although my work had been wonderful in many ways. To see so many different parts of the country; sometimes with the bonus of leaving a clothing or household textiles factory with a discounted product from the factory shop. It was some compensation for all the miles travelled, mostly on our ever-overcrowding motorway system. However, the challenges brought by the Big Bang (the sudden deregulation of financial markets in 1986) and competition from the rest of the world was slowly taking its toll on British businesses like textile manufacturing. Closures and other changes were starting to happen in Courtaulds.

In March, Poll Tax riots took place in several British towns and cities during protests about the Community Charge. The Charge was a system of taxation which provided for a single flat-rate, per-capita tax on every adult, at a rate set by the local authority. It was replaced by Council Tax in 1993.

Courtaulds plc demerged into two in 1990: its Textiles (yarn, fabric, and clothing) becoming one separate business, and Chemicals (fibre manufacturing and chemicals operations) becoming another. Their eventual successors had the less familiar names of Sara Lee and Akzo Nobel, both overseas registered companies. My final Health and Safety report to Courtaulds Textiles board members for 1989-90 was, nonetheless, a positive review of progress over the previous 5-years.

My line manager had seen the demerger as a chance to effectively demote me by bringing in another to do my job, but he was not open with me about his reasoning. Possibly it was because I was too assertive when some of my colleagues were threatened with redundancy. My decision to leave was immediate, though I kept that to myself whilst I searched for a suitable alternative position. It took me to several places. The MoD was one, and another was Ikeda-Hoover.

The latter (now Adient) was a joint venture between Ikeda Bussan of Japan, and Johnson Controls of USA to supply seating modules on a synchronous basis to Nissan's UK manufacturing facility in Sunderland. One day I received a call from the employment agency that dealt with Ikeda. It turned out that they also acted for Nissan, who were looking for a Safety Engineer. Later I discovered the reason: Nissan were commissioning a new low-pressure manufacturing plant at its Washington plant near Sunderland.

 On 9 September I had a pre-employment medical with Nissan's Occupational Health Service in Hebburn and on 18th, I accepted an offer of redundancy from Courtaulds. To 'help' me, I was offered a sham interview at Courtaulds Chemicals in Derby to demonstrate that a search for alternative employment had been made. On 24th, I attended the battery of psychometric tests (with specific reference to occupational skills) that was offered. The somewhat flattering results must have been embarrassing for my Courtaulds manager but would have been very favourable in Nissan's eyes.

The work I had done on developing machinery safety

standards with HSE's Textiles consultative group no doubt acted in my favour during interview at Nissan, especially in view of impending European Directives and International standards. I parted company on the Friday, deliberately absenting myself from a planned leaving presentation, to start at Nissan on the Monday 12 November. Ten days later the first cast of cylinder heads took place in the new casting shop, and further development of the shop was envisaged. Nissan bought and sold my Manchester flat, enabling me to buy a house in the north-east near Morpeth in Northumberland, without having the bother of having to secure a sale personally in Manchester.

1991

By November, expansion within the casting shop, and the introduction into production of the P10 Primera model the following summer, required a cross-functional team visit to Japan. Its role was to assess core-making and die casting machines at Sinto, a supplier in Osaka, and to see casting operations at Tochigi near Utsunomiya, Nissan's largest plant in Japan. Such teams performed a series of trials to ensure the capability of the plant and equipment being supplied met quality, safety and rate-of-production standards. A team usually comprised a senior production engineer who liaised with a Japanese counterpart as Trial Manager. He co-ordinated the work of the rest of the team which consisted of:
- A Production supervisor/member to consider line balancing, standard operations, and ease of work.
- A Mechanical member to review hydraulic and pneumatic systems and maintainability.

Nissan Sunderland safety team and 1993 Micra *(NMUK)*

- An Electrical member to examine control systems.
- A Safety Engineer (*Anzen Tojo*) to consider health, safe access/ guarding, and operator care.

Dave Thomas, then Senior Safety Engineer, strongly suggested that I should join the institute of Occupational Safety and Health (IOSH). I was admitted as a member in January and as a Chartered Member in 2005.

School and technical visitors at Nissan Plant, Tochigi

1992

In the next few years several visits took place to examine Fanuc robot cells and lines, to Hitachi and Komatsu in Osaka for transfer presses and to Fuji Engineering for dies. On one such visit, at night in Ota City, Gunma Prefecture (the home of the Japanese Subaru factory) I was awakened during the night in my violently shaking hotel bed. My first thought was that that other team members had entered my bedroom to play tricks on me. I was wrong: an earthquake had struck the skyscraper hotel, but it was built to withstand such an event.

In August, Micra production was started to replace the *Cherry*, but the UK slid into another recession in the early 1990s, and demand for steel declined. Ravenscraig steelworks closed, virtually ending steelmaking in Scotland.

Karakuri has always been practised to aid operators in the Japanese car industry. It is a Japanese word for devices that use old-school mechanical gadgetry – rather than electronics, hydraulics, or pneumatics – to accomplish a task. It is one of several concepts used throughout Japanese industry, which are worth briefly describing here:

> *Gemba Kanri* means a form of workshop management in which managers do a Gemba walk to see how operations are done in terms of machinery, materials, and the work environment, with the objective of improving quality and reducing cost, inventory, and lead times.
>
> *Kaizen* is a system to continuously improve operations in a gradual and methodical process through the involvement of employees.
>
> *Kanban* is a scheduling system for lean

The Kaizen process

Analyse it
↓
Develop solution
↓
Implement it
↓
Analyse result
↓
Standardise solution

(BJSTR)

manufacturing named after the cards or e cards that track production to avoid extra work in progress at any point of production,

Karakuri involves clever and inherently eco-friendly solutions to practical problems often involving hooks, pulleys, and counterweights to perform tasks. Carmakers use karakuri to transport objects over long distances or to ensure that the right parts for any given process are automatically delivered to workers, making tasks easier and more efficient. For engineers, dreaming up karakuri is a logical and creative exercise, and the devices can cut costs and improve productivity while simultaneously eliminating menial work and preventing mistakes.

A lot of attention was being paid to *Operator Care* – the Sunderland name for using kaizen to eliminate the unwanted aspects of *dirty, difficult and dangerous* tasks, whether physical or cognitive in nature. It had been recognised since the Bluebird launch that the anatomy of the Japanese male population differed from that of the UK. Therefore, before installation and use, adjustments needed to be made to the specifications of machinery and equipment coming from Japanese suppliers, and to conform with the new European health and safety legislation of 1992.

Advisers from Nissan's sister factories in Japan also attend Sunderland during the introduction of the equipment to produce new vehicles and their expertise was invaluable. Later, after the Renault-Nissan Alliance,

Aerial view of Tochigi plant with surrounding test track *(Nisssan Motor)*

Coremaking machines at Tochigi (Nissan Motor)

Sunderland also had access to French ergonomic methods. (*Méthode D'Analyse Ergonomique pour les activités répétitives*). In Sunderland a guide had been created to assess and improve manual handling operations in compliance with the 1992 Regulations which Safety Engineer Paul Douglass later computerised as an ease-of-use tool.

At first glance, investments in automation and robotics might suggest that the factory of the future will be the exclusive domain of machines, not people. In fact, nothing could be further from the truth: manufacturing is as reliant on human beings as ever, and technology will serve to enhance their work. Efficiency is achieved by preventing mistakes and maintaining quality by ensuring that workers are freed from monotonous tasks and reducing strain and fatigue from work.

Robotic manipulation of heavy cylinder heads for inspection (Nissan Motor)

Entire engineering teams are dedicated to studying ergonomics. For every process, they analyse the physical burden of certain actions (such as lifting, reaching, twisting, or crouching) or the mental burden of tasks that are repetitive, or require constant concentration. They then pick the most burdensome processes and come up with solutions to make life easier.

Certain assembly line processes are best suited for robots, particularly if they're simple and repetitive, yet relatively strenuous for humans. Industrial robots that work on things like welding and assembly are ordinarily kept in cages for safety reasons, due to their size, strength, and speed of movement. By contrast, cobots (short for 'collaborative robots') offer a perfect solution for processes where people and machines need to work closely together. Cobots are robotic arms with limited strength and speed of movement. In addition to being extremely nimble, they can be easily reprogrammed to learn new tasks. Fitting a head liner inside a car roof used to require an operator to lie on their back and push fit them in, using their feet. Today it is usually done by a cobot.

Body assembly robot welding line at Nissan, Sunderland (NMUK)

1993

Meanwhile, I became a member-appointed trustee of the company's UK Pension scheme and paid further visits to Yokohama engine plant, Oppama and Zama, now a development plant for the Micra, having produced millions of Sunnys and Cherrys. Weekend tourist visits were made, with that to the Landmark tower in Yokohama being certainly the most relevant, having experienced the Ota City earthquake. At 971 ft it was tallest building in Japan, comprising a flexible structure to absorb the force of earthquakes. Theoretically the structure resembles Japanese five-storey pagoda temples which never collapsed during a series of quakes.

1992 Micra

Katase-Enoshima beach, near City of Fujisawa was another tourist visit completed by an easy train ride from Yokohama. So too was the outward trip to attend the Japan Cup on the last Sunday in November. It is an international horse race over 2400m at Tokyo Racecourse in Fuchu. The return train trip was less easy. We misread the kanji font on the train's destination board and found out, too late after boarding, that it was a *kaitoku* train that did not stop in Yokohama, causing an unnecessary ride back. A trip to Oysuki, 80 km from Tokyo, in snowy

Tanzawa mountains, was altogether more pleasant.

In Briton Ferry, a second river crossing to carry the M4 motorway was completed.

The Railways Act of 1993 provided for the privatisation of British Rail and the introduction of a new structure for the rail industry, with many of the principal changes brought into effect on 1 April 1994. Two new statutory officers, the Rail Regulator, and the Franchising Director, were established to oversee the industry. *Railtrack* became a separate Government-owned company and was sold to the private sector in May 1996. British Rail was split into about 100 companies, almost all of which have been sold to the private sector or closed. The passenger services were divided into 25 separate units and sold to the private sector for periods of between seven and fifteen years. Other parts of the business including the freight operations and the rolling stock companies, are also in private hands.

Within a decade Railtrack was deemed to be a complete failure following three serious crashes at Southall, Ladbroke Grove and Hatfield in 1997, 1999 and 2000 respectively. The reports highlighted management failures with regards maintenance systems and human factors failures as major contributory factors.

An Osaka-bound Shinkansen at Utsonimaya

1994

Following the death of Dave Thomas, Sunderland's senior Safety Engineer, and his replacement by Barry Tupper, I picked up one of Dave's tasks which was to represent users on BSI's Low Pressure Diecasting safety standards committee. Work was already well underway when I joined so I can't say that I contributed much. Barry's move left a gap in the Body and Press shops, so I moved from Casting to cover them. Colin Adderley, who had accompanied me to Japan on a previous casting visit, took over coverage of casting.

Bodyshop supervisor Steve Greenwood was a keen practitioner of operator care, whose efforts got my support.

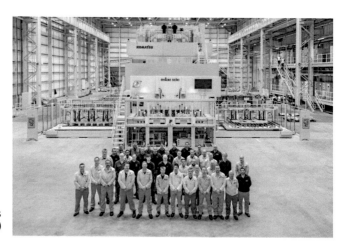

A 5,200-tonne press
(NMUK)

1995

One consequence of the 1994 reshuffle following Dave Thomas' death was that a further visit to Japan by a cross-functional team was necessary in March in preparation for the installation of a 5,000-tonne Hitachi Zosen transfer press to stamp full car body sides. The press ran well for twenty-five years. It was replaced in 2020 by a £52 million pound, 5400-tonne press to stamp up to 6.1 million panels a year for the new Qashquai and Juke models. The press includes a recycling system to segregate and process scrap, and the investment upgraded the existing blanking line to supply flat blanks (the sheets of metal that are ready to be pressed) to the XL press. All panels are pressed to pinpoint accuracy. This is a far cry from our 1995 visits for door hemming presses to Lamb's Engineering at Mildenhall.

A company benchmarking visit to Air Products at Wrexham resulted in encountering Stan Smith of Manchester Steel once again. Later I would revisit Air Products yet again with Amicus, the Union. My attempt to tutor evening classes at NORCAT in Ashington for students to obtain an IOSH Certificate was ill-advised. After a long working day in Sunderland and driving to Ashington, all on top of tutor preparation, it was so tiring that it unfortunately, simply had to be ended.

Victoria finished with Courtaulds and moved to Northumberland as Occupational Health Adviser to Northumbria Healthcare NHS Trust.

A 5,400-tonne press
(NMUK)

A break outdoors for the multi-disciplinary press team

1996

Victoria covered a large area for the Trust. Her role was to encourage good occupational health practice at all its operations. These extended, from Morpeth, north to Berwick and west to Hexham. As she did not see her post-employment future in the area, we eventually decided to buy, and then renovate, a house in Padstow, Cornwall.

To further develop my ergonomics and human factors knowledge for Nissan, I first attended a Biological Ergonomics module at the University of Surrey at Guildford. This covered topics such as biomechanics and anthropometry – essential ingredients for understanding human interactions with plant and machinery. When Nissan reviewed the course feedback it was decided that I should continue to attend later modules. Three were attended the following year: Ergonomics Methods, Practice and Management and Measuring Health and Illness. A fourth module at Nottingham in 1998, on Stress, was the last possible until I could return to Guildford for the Research module in 2001.

For some reason, the Company wished to reorganise the Pensions committee and I was asked to be reinterviewed as a member. I resigned in November and sought election to the Company Council, elected to represent Personnel, Engineering, Finance Purchasing and Information Systems. I worked day shift and, as a result, several urgent and interesting cases came my way, often at the request of the shift-working reps.

1997

At work on my 'day job', the subject of hand tool ergonomics came to the fore as the result of a *kaizen* project in the body assembly shop, initiated by operator Rob Seward. It involved improving the hand grip on spot welding guns which resulted in co-operation with the gun supplier, Obara, to amend the handle dimensions for the more comfortable use of UK operators. In parallel with this, work was being done with Sandvik/Bahco to better specify non-powered hand tools. The HSE were interested in this work and organised an industry seminar on *Hand Tool Ergonomics* at Sheffield in which I showed some of the changes that had been made to them at Nissan and why.

In the second Welsh devolution referendum, a majority of the Welsh electorate voted in favour of establishing a National Assembly for Wales by 50.3 per cent, on a 50.2 per cent turnout. Laws were subsequently passed to establish the National Assembly for Wales and grant it secondary legislative powers over areas such as agriculture, education and housing.

The Trussel Trust, founded in 1997, supported just two food banks in 2004, but an exponential rise after 2010 resulted in a hundred by 2011 and 1,300 in 2023. It supports a nationwide network of food banks to provide emergency food and support to people facing hardship. Its centres in the UK provide a minimum of three days' nutritionally balanced emergency food to people who have been referred, as well as support and advice to help people maximise their incomes and lift themselves out of poverty by tackling its root causes. Today, the Trust campaigns to end the need for food banks in the UK.

1998

The year started off as normally as others, but by its end the world was turning upside down for both industrial, but mainly personal, reasons. Industrially, the unprofitable ASW made a reverse take-over bid for profitable Sheerness Steel which was completed by April 1999.

When I attended the MSc Stress module at Nottingham University in March things seemed reasonably normal but, in the background, they were quite disquieting. My mother, who was not at all well in her care home was admitted to Neath hospital on 19 July and by the 27th, Victoria and I travelled overnight to see her. Victoria herself was concerned about a lump on her neck and was due to attend the Freeman Hospital in early August for biopsy. Dr Meikle promised to phone at mid-day on 22nd with a good/bad result from the biopsy. It was bad.

My mother died on 1 October with my brother Andrew attending to my mother's arrangements. Victoria was back and fore as an outpatient to Newcastle General where a first CT (computed tomography) scan in December.

During these events the Good Friday Agreement was approved by voters in two referendums in Northern Ireland and Eire on 22 May to come into force in December the following year.

The Government of Wales Act created a National Assembly. Its complement of 60 members included 24 females.

Ludlow 21, a local voluntary group, was formed to promote sustainable living in the Ludlow area within a fair and just global society. It is an independent body with no political or council ties, but is a local voice for sustainability, with very active groups that focus on fair trade, energy, transport, education, food, wildlife and more.

1999

Cobots at Nissan – used where staff and cobots can work safely in close proximity

On 27 March, the Nissan-Renault Alliance was formed to achieve maximum efficiency by combining the strengths of both companies. It aimed to develop such synergies though common organizations, cross-company teams, shared platforms, and components. Almera model production started in Sunderland and continued until 2005, when the Note and Qashqai models were introduced to replace it. Cobot arms were appearing more frequently. At Yokohama, for example, they were being used to loosen bolts on cylinder head cam brackets and for engine block intake manifolds, the latter having been cast at Sunderland since 1991.

In the rest of the world the, now private, British Steel merged with the Dutch company, Hoogovens, to form *Corus* Group to create a UK steel workforce of 28,900 at the time of merger.

Prince Charles spoke in Welsh to the sixty new Assembly members and the Queen signed a special edition of The Government of Wales Act to symbolise the transfer of powers from Westminster to Wales.

I was oblivious to many such happenings because, from January, Victoria had been undergoing her sixth cycle of chemotherapy at Newcastle General as an outpatient. By the end of July, she was no longer an outpatient but an inpatient at the Freeman, the RVI, and the General in turn. In July, an aggressive lymphoma in her chest and abdomen had been detected. Doctors were concerned at the speed of her relapse. She was in a high-risk category and then her spine was fractured, so a hip operation was carried out. By 8 September she had spent 43 consecutive days in one or other of Newcastle's three major hospitals. Her brother, Adrian, and mother, Frances, came to visit, and on 23rd Victoria and I were married in the RVI by Registrar General's Licence. On 18 October she was discharged for palliative care for MacMillan nursing at our, now, rented home in The Glebe, Stannington. She had to be re-admitted to the RVI where she died on the 28th, after a total of 86 days in the three hospitals. The saddest part was that she pre-deceased her mother.

On 15 December I returned to work for the Company Council's salary negotiations. Nissan had been extremely generous in recognising my need to be absent from work and I owed it to everyone to return.

An impromptu musical event in Vinales, Cuba

2000

St Edmund's House Padstow

Rick Stein, known as 'the TV chef', bought 6 St Edmund's Lane Padstow from me in September 2000. He converted it into St Edmund's House with 'six luxurious and peaceful guest rooms with its own beautiful gardens in the heart of Padstow behind the Seafood Restaurant'. One can still stay there for bed and breakfast at £330 a night double!

That month some further awful news came from Victoria's family: brother Adrian's wife, Janet was diagnosed with motor neurone disease. She died that August.

After moving to Stannington there was a renewed proposal for a nearby building development that the developers, Northumberland County Council, had previously denied would take place. We decided to mount a campaign which involved posting 'NO' notices on the fences along the A1. A BBC producer saw them, and I ended up in the Radio Newcastle studios on several early mornings before work at Nissan to be interviewed about the protest. A strange coincidence then occurred. Sylvia rarely listened to Radio Newcastle, but on one of those mornings she had the radio on and heard me. It got us together again and in December we met in Havana to explore as much of Cuba as time permitted.

Later, in 2000, Nissan colleagues on the Company Council persuaded me to become a Union Rep and not just a Company Council Rep. This started my stronger involvement with the Amalgamated Engineering and

Electrical Union (AEEU). Soon afterwards I attended an AEEU Workplace Reps Course at Gateshead. In my day job, my advocacy for the separation of works traffic from materials handling safety was successful. In recognition of the different types of risks involved in *just-in-time* production methods, the creation of a separate Traffic Safety Committee was confirmed.

The plant to be allocated for the new *Micra* model had not been announced. It was customary for the company to encourage competition between plants for new model allocation. In this case it was to be Sunderland or Flins (near Paris), but when the Company Council from the Renault plant visited Sunderland their union reps quietly informed the local reps 'not to worry' because they certainly did not want it at Flins at that point. Today, Flins produce a later model Micra.

I had purchased my brother Andrew's share of Ruskin Street and, in December, with all the necessary work completed, it became available for occupation again.

2001

Early in 2001, I moved to Sylvia's home at Railway Cottages in Durham until we completed the move to Ferryside in south-west Wales. By attending AEEU Region 3 Conference at County Hall in Durham in January, I realised that I would likely be a delegate to the National conference at Blackpool in June. Ian Davies, then full-time Union Officer for Nissan, Sunderland, expressed his disappointment that I was leaving the company. He suggested that something might be available on a self-employed basis in the AEEU education department. As it turned out, the AEEU merged shortly afterwards with the MSF (Manufacturing, Science and Finance) Union to form Amicus.

My last day at Nissan followed on 9 February and the following day I said goodbye to my colleagues when we assembled at Fairyhouse races near Dublin. It rained all day! On return, Sylvia and I made our first joint visit to the Ferraros in a much brighter and sunnier Rome.

In April, Messrs Galsworthy, Radford and Hutin, three

A farewell to Nissan with Sunderland Health and safety team

Farewell at Fairyhouse with Barry Tupper

blast furnace workers on Corus' number 5 furnace at Port Talbot were killed in a catastrophic explosion. There was a loss of containment of 200 tonnes of furnace contents with the whole furnace and contents being raised vertically by 0.75m due to the violent release of energy when cooling water encountered the contents. Corus was fined £1.33 million with £1.74 million costs. Kevin Downey, now a supervisor, was prominent in recovery, but tragedy awaited him, too, five years later.

During an Employment Tribunal at Newcastle, Ian Davies confirmed to me that Mike McCartney, AEEU's Education Manager, would like to see me. At Esher Place in July, he later confirmed that I could join the team of tutors which delivered the union's education programme. Sylvia encouraged me keep active during my 'retirement', so the offer fitted in well for that reason. We finished that trip to Esher and London with the good fortune to visit both the Cabinet War Rooms and the House of Commons, where Ian Paisley senior was holding forth.

In August, the purchase of *Llwyn yr Eos* in Ferryside was completed, but on the day that we took possession, we discovered a planning application for forty houses in the field behind the house. This gave two causes of concern; firstly, flocks of curlews frequented the field and, secondly, concern at the impact of water run-off into an already flood-experiencing low-lying area of housing below.

Esher Place *(Unite)*

Ironically, the development was to be called *Parc-y-ffynon* (the field with the well).

Early one balmy September evening, whilst following a more circuitous route from Ferryside to Briton Ferry than normal, I was confused by reports on the car radio. At first, I thought it was a piece regarding the ongoing investigation of the April explosion at Port Talbot. Eventually, I understood that the report was referring to the suicide attacks by Al-Qaeda on the eastern USA which killed 2,977 people, making them the deadliest terrorist attack in history, and instigating the multi-decade global war on terror, fought in Afghanistan, Iraq, and elsewhere.

The first course I delivered for the Union came at Wakefield in October, where I stepped in to complete a course for Mick Reed. Other visits came thick and fast, ranging from the main residential training centre at Esher Place, to factory visits at BOC in Shoreham, Unipres in Washington, Crewe Loco Works, Honeywell in Glasgow, Air Products at Wrexham, and Belfast. For good measure I presented the last course at the Union's Cudham residential training centre. Somehow, I also fitted in an important visit to the AA Headquarters in Basingstoke as part of research for the project for my MSc award.

2002

It was an eventful year all round. I was the first to move to Ferryside in January to look after the rebuilding of Llwyn yr Eos, with Sylvia following in July. For part of that time, we absented ourselves to Australia, but lived afterwards in Ferryside for five eventful years.

Sometime after co-option to the local Community Council, I was approached by the Clerk, Alan Dark, to consider the situation at Ferryside Rugby Club which was close to our Llwyn yr Eos home.

He, Dr Roger Griffiths and Vernon Thomas, a solicitor, were concerned about the future of the club in view of its financial and position and management. They were looking for support to help secure its future. In the first instance the problem was to repay an outstanding bank loan.

The chair had resigned, and the steward had been removed, suggesting that the position of the club was hopeless. I was informed that more effort could have been made to manage the club better; in fact, there was little incentive to do so because of the value of the land on which the clubhouse stood. After I accepted the chairmanship, I began to realise exactly what was involved. The rugby playing side of the club was in decline for several reasons. A major financial drain was also taking place because support for the club had halved as the village was split in two factions regarding its future. This was a sad, uncertain, and unnecessary situation.

Some matters were helping to offset the financial drain; others were not. The presence of itinerant cockle-pickers both helped and hindered, whilst the introduction of digital television, and the use of the clubhouse it entailed, and as a filming HQ, all helped the club's footfall. Then

Ferryside Fisherman sculpture

these matters all became compounded by the effect of the Licensing Act 2003, which became a story in itself.

In December, at fifty-seven years of age, I was awarded an MSc at Guildford Cathedral and, a week later, Sylvia of similar age, received her doctorate in Durham Cathedral. Her research concerned *Intercultural Competencies of Upper Secondary learners of French*. In contrast I became a Member of Ergonomics and Human Factors Society and a Chartered Member in 2005. Shortly afterward, on the advice of an old friend from the steelworks, I went to see Roger Derrick, another former steel-works colleague, at his Holistic Services Consultancy at Llandarcy, near Neath. He offered some consultancy work for the Welsh Government which would be on top of the tuition work being done for AEEU/Amicus, Ruskin, and Northern Colleges.

In a bid to stem losses, Corus closed Ebbw Vale steelworks; Allied Steel and Wire (ASW) formed by British Steel and GKN went into receivership, closing Sheerness for first time and its Cardiff works, for a while, until Celsa of Spain bought it the following year.

The Welsh Government separated its executive and legislative functions.

2003

The Licensing Act 2003 was introduced. Broadly speaking, licencing is about premises, events, and alcohol. It has four objectives: the prevention of crime and disorder; public safety; the prevention of public nuisance; and the protection of children from harm. In Ferryside, the Act was seen, two years later, to be a malicious opportunity for the Rugby club's opponents to use the Act to harm the Club.

Meanwhile, trips to France and Ireland and two visits to Italy were taking up some time, but the main time demand came from some sixteen week-long courses at venues as far away as Needham Market and Tynemouth. There, at the Little Haven Hotel, near to where the Great North run ends, my favourite pastime was to watch the car carriers entering and leaving Tyne Dock and the passenger/car ferries sailing to places such as Ijmuiden/Amsterdam. The rest of my timetable comprised some thirty, one-day, more local, visits to such varied establishments as the DVLA in Swansea and Smurfit printing and packaging in Weston-super-Mare. Ironically, I spent less time driving than before or afterwards.

Car carrier
Port of Tyne

2004

Ferryside Flood Awareness Group persuaded the local Assembly Members to support our campaign to mitigate the flooding risks of the housing development at Parc-y-ffynon. We succeeded to the point of obliging the developers to install hydro-brakes to capture and delay run-off from the fields and hills above.

Whilst delivering a residential course at Plymouth, I received a call from Sylvia who was in tears when she met me off the train at Ferryside station. Grandson Felix Ferraro had just been diagnosed with leukaemia. One small consolation was that the builders had started work on getting our house ready at Via Marsala in Poggio Mirteto which would enable more frequent visits in addition to the thirteen we had made in the previous four years.

Vacone – a Comune in Sabina, Italy

In December, BBC Digital chose Ferryside and Llansteffan to conduct its first switchover trial to phase out the analogue signal to terrestrial digital television. The rugby clubhouse became the BBCs trial HQ. There, the 350 homes that had received free Digiboxes from the BBC could receive advice on how to use them. The use of the club as the digital HQ reminded people in the village to use the club.

The UNIFI and GPMU unions merged quietly with Amicus in 2004.

2005

Twelve years previously, the unregulated Three Rivers Fishing area of Carmarthen Bay had witnessed the *Cockle Wars* in which hundreds of cockle pickers had descended on the area. Once again, in 2005, a bumper crop of cockles led to the opening of beds for commercial harvesting on the Towy estuary for the first time in four years. Cockle pickers, hoping to earn up to £500 a day, had travelled from as far as Scotland.

After a hard day on the cockle beds, the Rugby club was a sanctuary for the pickers to obtain refreshment which provided a welcome income source for the club. Although, at times the club became a little noisier than normal, it led to some exaggerated complaints from those neighbours who wished ill of the club.

Following a holiday visit to Canada, where I celebrated my 60th birthday, Sylvia and I were returning from Rome for a late evening flight from Ciampino to Bristol by EasyJet flight 642. We were abruptly informed that the flight had been cancelled. It was the second successive day

Poggio Mirteto
centro historico

that EasyJet had cancelled the flight, but travellers considered the cancellations to be for reasons within the operator's control. EasyJet had been criticised in Germany for not observing European Union law on compensation (and assistance to passengers) in cases of denied boarding, delays or cancellations (Regulation 261/2004). When flights are cancelled, passengers are supposed to be reimbursed within one week. In 2006, the airline did not always refund tickets in a timely fashion and passengers occasionally had to wait longer for reimbursement of their expenses.

There were some large young families at Ciampino Airport that night, for whom all the uncertainty about the flight was particularly harrowing. Passengers were provided with all sorts of nonsensical reasons, such as holes in the runway for the cancelled flight, yet planes were still using it. Eventually, at 3 am, travellers were bussed to a hotel at Ostia where a complete meal with wine was laid out in the dining room. But at 3.30 am?

It did not end there. To return us to Bristol, EasyJet required us to pay for an Easy Jet flight from Rome to Newcastle with an onward fight to Bristol. On complaining to the company, on our return home, we were offered £62.48 each, minus £5. This amount did not comply with EU flight compensation rules which had been introduced in February the previous year.

EasyJet had failed to meet its responsibilities, so we decided to file a claim at our local small claims court in Carmarthen. Initially, Easy Jet said that they wanted to settle out of court, but they never made an offer, so we went ahead.

That October, at County Hall Carmarthen, the Licensing Committee heard the allegations that the Rugby Club were in breach of the Licensing Act because it had failed to prevent public nuisance. The untrue allegation concerned disturbing noise on a specific date and time. In defence, the Rugby Club's representative, Vernon Thomas, was confidently able to state, that at the time of the alleged offence, the police were at the club investigating a previous break-in. The allegations were refuted, the claimants were discredited, and the License was renewed.

When the Ergonomics and Human Factors Society upgraded my Registered Membership to Chartered, after approving the evidence I submitted on how I had applied ergonomic principles, it was time to close the cover of a busy year.

Poggio Mirteto, near Rome, became such a well-visited venue for family reasons that we decided, whenever possible to combine these visits with trips to other places. In Italy these would typically be in Lazio, Tuscany, Emilia Romagna or Basilicata, but sometimes, too, to the Gers in south-west France, or even Andalucia in Spain.

A hilltop village in Pollini, Basilicata, Italy

2006

The next year turned out to be probably the busiest ever for all sorts of reasons, not just because of the work of a freelance consultant.

Firstly, we were informed that our court claim against EasyJet, for a total of £854, was successful because they had not contested the claim. On 4 February a Guardian article by Miles Brignall reported that the case was significant because it paved the way for passengers to claim compensation for cancelled fights. Later that month BBC News24 reported on the case. I was in Sheffield when I received a call to be interviewed live in Manchester. I was not at my best, but it kept the matter in the public eye and the *Guardian* continued to report on the matter.

The claim against EasyJet, for a total of £854, was successful

Tragedy struck on 25 April when Kevin Downey, who was previously referred to in 1973 and 2001, died at the age of 49 at Blast Furnace 4. He died in hospital after falling into an uncovered slag runner when his vision was impaired by steam. Corus were fined £500,000. This horrific incident no doubt influenced the plot of Michael Sheen's 2024 production of *The Way*.

115

Flight path: Air users can take their lead from separate court cases which saw Philip Adams (above) see off easyJet while Thomas Cook (left) had to compensate David Harbord

them to the plane and put them up in a hotel while it was being repaired."

While Mr Harbord is still waiting to receive his money, Philip Adams and Sylvia Duffy of Ferryside, Carmarthen already have theirs. The pair, who regularly travel to Italy, were due to fly from Rome to Bristol when their easyJet flight was cancelled – coincidentally, this was also on August 5 last year.

Instead they were offered a flight to Newcastle, but told by easyJet they would have to pay for the connecting flight back to Bristol. "The airport was in chaos, not least because this was the second day that easyJet had cancelled the flight. One young woman had been waiting 30 hours, and there was a family with eight children, not knowing what to do," says Philip.

When they eventually arrived home and complained, they were offered the cost of their original tickets (£62.48 each) minus £5, but denied the compensation set out in the regulations.

Philip filed a claim with his local small claims court. "Prior to the case easyJet rang me with a view to settling it out of court. I rang back and left a message but didn't hear anything. Last week I heard that it had been uncontested – and we had effectively won."

The pair claimed a total of £854 – including the cost of the initial tickets, the extra flights, a taxi ride and compensation of €400 per ticket.

"easyJet refused to own up to their responsibilities, which was why I wanted to take it to court. We were quite lucky in that we got back home – some of the other people due to fly back that day will have endured a nightmare. Looking back, we should have all swapped phone numbers and fought together; I suspect that most will have let the matter drop, unaware they are entitled to compensation," says Philip.

This view is shared by David Harbord. "This case raises some important issues – not least, it shows the Air Transport Users Council (AUC) to be completely useless. Once I received the judgment, I rang the AUC to ask whether they will be contacting the other passengers to tell them that they are also entitled to compensation, but they weren't interested. The Civil Aviation Authority were the same. The fact that the regulatory authorities are so useless is a disgrace," he says.

● Were you on the same flights as Messrs Adams and Harbord? Mr Adams was due to fly on EZY 642 from Rome Ciampino to Bristol. Mr Habord was due to fly on TCX 24K from Stansted to Váncouver via Manchester. Both flights were on August 5 2005. If you were, or were on another flight cancelled in similar circumstances, contact the airline. If it refuses, dust down your legal briefs and go to the small claims court.

m.brignall@guardian.co.uk

EasyJet *Guardian* **report**

That was not the end of a year of litigation, with the next event coming totally out-of-the-blue. I was asked to meet a Reed Smith barrister at Temple Bar in September to give evidence about an employer liability claim for noise-induced hearing loss. It was made by a former Courtaulds employee who worked at Meridian Fabrics in Sutton in Ashfield from 1971 to 1989.

The case was heard at Nottingham High Court from 26–30 October. Dr Peter Cooper and Brian Arthurs were called as witnesses to defend the company. Thankfully I was not called. The case turned on the date of knowledge that noise was known to be harmful. The judge found for the defence. Witnesses showed that where noise levels potentially exceeded certain levels, the company had taken any legally required action since the early 1960s to prevent noise which might lead to hearing loss as well as comply with the stricter standards of the 1990s.

I was not aware, until I wrote this piece, that the case later went from Nottingham, with others, as a test case to the Court of Appeal. Even more surprisingly, it then went to the Supreme Court as *Baker v Quantum Clothing* (2011). The Supreme Court judgement stated that:

> *Courtaulds and Pretty Polly, however, were in a special position. By the beginning of 1983 they understood the risk that some workers would suffer damage from exposure to between 85 and 90dB(A)lepd, which distinguished their position from that of the average employer. Allowing a further two years to implement protective measures, they were potentially liable at common law from the beginning of 1985.*

However, Mrs Baker's claim was dismissed on the different basis that her employers had not committed any breach of common law or statutory duty. That meant that the evidence from Messrs Cooper and Arthurs was sufficient defence.

For Wales the Government of Wales Act separated the Government, which executes policies, from the National Assembly which holds the government to account.

Italy v Wales at Flaminio Stadium Rome

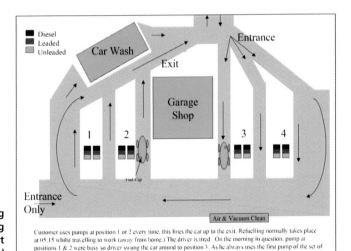

Where misfuelling occurs: a filling station forecourt *(Paul Douglass)*

118

2007

The previously mentioned visit to Basingstoke in 2001 was to collect information from the AA about the misfuelling of light vehicles. My motivation to do it was after I had fuelled up with diesel instead of petrol, on my first day at work after Victoria's funeral. I had wanted to know why I had done it. and that was my motivation for the research. My tutor was keen to get the thesis published. *Light vehicle fuelling errors in the UK: The nature of the problem, its consequences and prevention* duly appeared in Applied Ergonomics, under my name and that of my tutor, Geoff David of the University of Surrey.

The research involved a lot more than interviews. Even during holiday visits to Canada and Australia, it was impossible to resist examining filling station forecourts and driver behaviour whilst refuelling, as shown on pages 118 and 122.

Sylvia was quite tolerant about my misfuelling antics but uneasy about the commitment required at Ferryside RFC, even though her rugby experience was broadened by attending Italy v Wales at Flaminio Stadium in Rome. Village life in Ferryside, which was ten miles from Carmarthen, the nearest town, was not working for her. She was disappointed with the tuition received in her attempt to learn Welsh, but we did receive twinning guests from Lesneven in Brittany. Together with consideration of her sister's serious ill health in Manchester and the work she was doing for the Open University, which was taking her to the Marches, she suggested that we move to a 'halfway house'. The situation led us to Ludlow.

In the steelmaking world, Tata Industries, the Indian

company bought Corus and its 23,000 numbered work force had a new employer.

By this time, I had prepared and delivered for Amicus, a course on *Ergonomics for Workplace Representatives* which I hoped would receive a widespread reception. In the trade union world. However, Amicus merged with the Transport and General Workers' Union to form *Unite*. The new General Secretary seemed to favour the TGWU way of delivering the education programme over Amicus. Several of the freelance Union HSE tutors (The Consortium) met at Oxford to consider the situation.

During a welcome visit to Strasbourg we took an opportunity to see the official seat of the European Parliament. It meets, and all votes its must take place there, twelve times a year, although the majority of work takes place in Brussels.

Banks such as Northern Rock had developed a new business model. Instead of raising deposits and then lending them to house buyers, banks originated mortgages and the distributed them by selling cash flows coming in from mortgage repayments (securitisation). In 2007 the market froze because of fears that these had been sold to people with poor credit rating histories and who would be unable to repay the loans. Northern Rock was soon in trouble because only 25% of its mortgages were funded by traditional deposits.

Following the deterioration in solvency within Northern Rock in September 2007, it was nationalised the next February.

2008

In March, we sold *Llwyn yr Eos* and moved to Old Street in Ludlow after completing a tranche of improvement work there. After a while, an interesting Grade II listed house became available on Corve Street, but it also required even more work than Old Street to reach our requirements. Following a structural engineer and surveyor's advice and much building activity, that work was completed the following August. Fortunately, these moves could take place without mortgage worries.

Dinham Bridge, Ludlow
(John Fleming)

For the Union, I had developed course materials for Managing Stress and Ergonomics and made several visits to Honda Manufacturing in Swindon for delivering both *Environmental* and *Representative* training. There, many south Walians, displaced from the pits and steelworks found employment. Many, including shift workers, travelled daily from as far afield as Ebbw Vale. Honda found in Swindon what Nissan had found in Washington, an experienced industrial workforce, albeit from the railway workshops rather than the pits and shipyards, and a suitable site in the form of a redundant RAF airfield.

Honda was to close its operations in Swindon completely by July 2021.

Other residential centres used by clients for courses were at Eastbourne, Quorn, Dumbleton Hall (Evesham) and Wentworth Castle (Barnsley). At other times less salubrious venues were used, such as a scrapyard in Newport and a cricket pavilion in Faversham (Kent), all imaginatively dealt with by a new Member of the Institute for Learning.

Meanwhile, the financial crisis had worsened. In September. Lloyds-TSB rescued HSBOS and Bradford and Bingley were nationalised. A month later. the Government bailed out RBS, Lloyds-TSB and HSBOS.

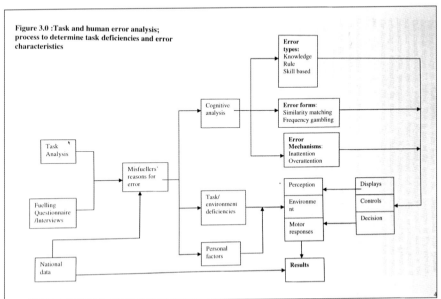

Process to determine error characteristics when fuelling a car

2009

In April Chancellor Alastair Darling revealed that bailing out the banking system would result in the largest budget deficit in UK history.

After the August move to Corve Street we met our then neighbours, the Collinses of Chester, who owned number 143 as a holiday home. Peter Collins kindly gave us a copy of *The History of a House* – a booklet that he had encouraged local historian David Lloyd to prepare on 143 and 144 Corve Street, which were originally one house in the early 19th century. He was also in the process of writing *The Art of Needlemaking*, which he later published.

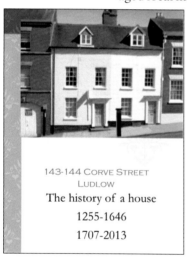

143-144 CORVE STREET
LUDLOW
The history of a house
1255-1646
1707-2013

Peter Collins inspired this booklet

Once again, working for the Welsh Development Agency's Holistic Services, I encountered low pressure aluminium diecasting for automotive products at Contech in Welshpool. Many trips were taken to Esher by car and train but when the Wrexham and Shropshire Railway offered a service to Marylebone with old-fashioned catering, it made Esher and London trips such a pleasure.

Waterford-Wedgwood went into administration and Woolworth's finished trading. A 50% tax rate was introduced for those earning over £150,000. A scandal arose over MPs' expenses and, with unemployment at a 14-year high, Lindsay Oil Refinery workers struck against the use of foreign workers.

Computer Weekly published a story about defects in the Horizon computer system which Fujitsu was managing for the Post Office. It would take eleven years before

Prime Minister Johnson committed to hold an Independent Enquiry into why hundreds of sub-postmasters were wrongly charged or imprisoned for accounting offences because of the defects in Horizon. It became a Statutory Enquiry in June 2021 which also investigated cover-ups by Post Office management and their lawyers.

In Port Talbot, two years after Tata acquired the works from Corus, Blast Furnace 5 produced its weekly record of 57,249 tonnes – an output never to be exceeded.

Spain 2013

2010

In the 2010s, global warming became increasingly noticeable through new record temperatures in different occurrences and extreme weather events on all continents. The CO_2 concentration rose from 390 to 410 PPM over the decade. Combating pollution and climate change continued to be areas of major concern, as protests, initiatives, and legislation garnered substantial media attention. The Paris Agreement was adopted in 2015.

Although Britain had four prime ministers, Brown, Cameron, May and Johnson, it was mainly Cameron and May for most of the decade, and it was under them that it became a decade of austerity and food banks, following the financial banking crisis of the preceding few years.

Work with Unite was still forthcoming, particularly for delivering courses in the south-east at places such as Crawley, Faversham, Maidstone, and Sittingbourne. Involvement with security staff from Gatwick Airport was particularly interesting. At Lincoln, I enjoyed delivering a Preparation for Retirement course, brushing up and adding to my own knowledge from my time as a Member Trustee of the Nissan Pension Plan! In addition, I enrolled for a University of Bath distance course for the IEMA-approved Associateship module to fill any unexpected gaps in my time and knowledge.

Despite all these other occupations, Sylvia and I managed to reach Calabria to see the Ferraro family. Lamezia Terme, near Catanzaro in Calabria was our destination, and the overnight sleeper train from Milan was our means of getting there. The thermal baths at Sambiase resulted in the suffix *Terme* being added to Lamezia. Today, the industry in the area comprises the manufacture of red clay pottery and the production of mediterranean crops such as olives, grapes, chestnuts and, above all, Calabrian clementines. Previously, the Nicastro area had mulberry trees and silk production, but they declined, and the local population with them, by the 1940s.

In return we joined up with the Ferraro family in Cornwall to cycle the Camel Trail, see Tintagel and Fowey whilst enjoying a visit to the changing Padstow into the bargain.

When the Staffordshire Hoard went on public display at Hanley, it was imperative to head in the opposite direction to join the long queues to see the treasures for fear that they would forever disappear to London or elsewhere.

Catanzaro ware

Alhambra Gardens

2011

Ironbridge B Power
Station in 2007
(Wikipedia)

E.On UK, one of the big six UK energy suppliers was one of Tern Associates' clients. It was on E.On UK's behalf that I visited Ironbridge B coal/biomass fired power station a few times. The station first fed to grid in 1969, but the future of the station was not seriously in question at the time of my visits, but it soon was. That is why I was grateful to be able to visit when I did. Ironbridge B stopped generating electricity in November 2015, with the decommissioning process continuing into 2017 and the twenty-seven-month demolition process beginning in December 2019, starting with the four cooling towers. Impressive conservation measures had been put in place at the plant, to ensure that the station merged as far as possible into its natural surroundings, unique amongst British coal-fired stations. When viewed from Ironbridge, the surroundings of the station were hidden by wooded hills. The cooling towers were deliberately constructed using concrete to which a red pigment had been added, to blend with the colour of the local soil. The towers could not be seen at all from the world-famous landmark, The Iron Bridge. The station's single chimney was Shropshire's tallest structure, being higher than Blackpool Tower or London's BT Tower.

In nearby Wales, from Berriew, on the Montgomery Canal, we sometimes unfolded our bikes to cycle along the towpaths to reach Newtown in one direction and Welshpool in the other. Despite the canal's proximity to the parallel A483 road, the canal is quite isolated from that road's noise. Making use of one of the pubs, where the old stone bridges cross the canal, it is a pleasure to watch the swans. Gradually, the canal is being restored where it has fallen into disuse so that recreational boats can make better use of it. Wildlife thrives along the canal. It's one of the most important canals in the country for nature, much of it is a Site of Special Scientific Interest and the Welsh section is of international importance, designated a Special Area of Conservation for its aquatic plants.

The Canal was devised with a different purpose from most other canals of the time. Whereas other canals could generate sufficient revenue from cargo carrying to be financially viable, the Montgomeryshire was planned to serve a more rural area which would not offer such opportunities. Instead, the primary purpose of the canal was to transport lime for agricultural purposes which would allow the Upper Severn Valley to become better agricultural land.

The third referendum saw voters support full primary law-making powers for the National Assembly over specified areas of governance. The National Assembly was renamed 'Senedd Cymru' (in Welsh) and the 'Welsh Parliament' (in English) (also collectively referred to as the 'Senedd'), which was seen as a better reflection of the body's expanded legislative powers.

The Welsh Labour Party advocates for further Welsh devolution and 'far-reaching federalism; with powers equal to those of Scotland and Northern Ireland, whilst the Welsh nationalist party, Plaid Cymru has described devolution as a stepping stone towards full Welsh independence.

Joyce and Derek Ridings

Joyce Ridings

I first met Sylvia's sister, Joyce and her husband Derek just before Christmas ten years previously when they were about to move from Glossop to a larger home near Whaley Bridge which could accommodate a design studio.

Joyce was both an imaginative clothes designer and earnest manufacturer who traded both under her own brand name and manufactured under contract for Jaeger. It had been her ambition to do so since she attended Manchester Art College, and it was Derek who provided her with the essential practical support as the business grew. Gradually, it comprised both factory operations and retail activity at premises at home and overseas.

Only a small portion of the work was contracted out because the philosophy of the limited liability company was that manufacturing should take place in the UK, employing direct labour, and very much operating on ethical principles. These were welcomed by corporate clients like House of Fraser and John Lewis and private customers, whether they were being sold under the Joyce Ridings brand or of others, like Jaeger.

Joyce was extremely enthusiastic in her work and my previous experience with Courtaulds enabled us to hold many interesting conversations about manufacturing matters. One such topic was the offshoring of manufacturing that had led to the eventual demise of Courtaulds Textiles. Clearly the resulting competitiveness from cheaper overseas manufacturing had the potential to pose problems for a business operating on such a principled philosophy as Joyce's.

The list of defunct retail companies had already started to grow since before the Millenium. Wm McIlroy, formerly of Courtaulds, went in 1998. The conversion of Great Universal Stores into Experian Finance in 2006 said a lot about the relationship between manufacturing, retail and finance. Other well-known names such as BHS, Ethel Austin and Debenhams were to follow. Fast-fashion clothing and online buying from the likes of Zara, H & M, and Topshop were replacing the established household names.

Joyce was chosen by Manchester City Art Galleries to display a retrospective exhibition of her work at The Gallery of Costume in Platt Hall from May until September. Its aim was to 'document the timeless elegance of her designs from 1966 to 2011 as well as showing her steadfast devotion to Manchester'.

Eventually, some major customers forsook Joyce Ridings' products in favour of cheaper overseas suppliers. With better succession planning it might have been possible for Joyce to attenuate such negative trends on both her health and the business. Nevertheless, it was the effects of the 2008 financial crisis and the austerity that followed that undermined the Ridings' business. In 2013 Joyce's ill health would take its final toll and it was left to Derek to wind up the business before he died two years later.

One of her obituaries in the *Draper* by colleague Angela Giles said much of Joyce and her approach to work:

> 'With dreams of becoming a fashion designer, it was an inspiring art teacher who encouraged her to go to the Manchester College of Art and Design to study fashion.
>
> Fresh out of college and full of enthusiasm, Joyce and a fellow student conceived the label *Qui*. Designs were eagerly welcomed in local shops and boutiques on the King's Road London, the epicentre of the swinging '60s. The business began in a tiny workroom in the centre of Manchester, later moving to larger premises as success grew.
>
> Fashion became more cosmopolitan in the '70s

and the European inspiration behind Joyce's designs helped increase sales rapidly. The quality of manufacture was consistent, with a growing team of highly skilled staff following Joyce's belief in the highest level of craftsmanship.

The label *Joyce Ridings* began showing at the newly created London Fashion Week in the '80s making the name accessible to a wider audience. She also took her collections to Paris and New York, further growing her reputation.

By the '90s, as well as her own collections, Joyce had a close association with Jaeger and her designs were sold in their flagship stores in the UK, America and Japan.'

A Joyce Ridings design

2012

The Consortium of Unite Tutors held a final meeting at Kidderminster, but nothing was resolved about creating a consultancy of that name. Perhaps that was just as well because of the situation with Sylvia's sister and brother-in-law's health at Greens Deep. It was prudent to focus attention there, along with pursuits that could be postponed or abandoned should that be necessary. Three such pursuit did emerge: to restore a longcase clock, to work at the Citizens' Advice Bureau (CAB) and to research family history.

Perhaps the family history turned out to be the most rewarding. It started off in June because of questions about a Widow's Penny found in our garden shed at Ruskin Street with the name of Lee Adams on it. We had no idea who he was or why the medal was there. Field work in Pembrokeshire, at New Moat and Walton East followed, to find out more about the Adams side of the family. It became clear that the advent of the railways to west Wales enabled family members to leave agriculture for an industrial life further east. Although the work was 'completed' and printed in a full-colour booklet in January 2013, further information became available which involved some corrections and additions.

Although the Citizens Advice Bureau had a Ludlow branch, I thought it advisable to apply elsewhere. After an interview and familiarisation courses I became a Gateway Assessor, spending an interesting half-day per week at the CAB in Leominster. I was originally interested in consumer work but soon realised that CAB covered a lot more than that. Cases on benefits, work, debt and money, health, housing, immigration, and the courts all came through the door sooner or later. Even at this early stage

Olympics come to Ludlow, 2012

of a decade of doom I was issuing vouchers for Leominster's food banks. What I did not realise was that sometimes these were being tendered for items such as baby food and nappies. The country was now on a regressive trajectory.

A friend who lived near Swansea had found a clock face and mechanism in an old barn and offered it to me for restoration, instead of throwing it out. That became my third project.

Swansea clock

2013

History was being made in Sunderland in 2013. Although imported Nissan Leafs had appeared in the UK from March 2011, the first electric car made in the country by

Nissan Leaf
VO20YMK

Nissan first rolled off the line in 2013, with a third-generation model now scheduled for 2026. Envision/ AESC's adjacent giga battery factory supplied the Leaf production from its own opening in 2012 and expects its 12 GWh additional factory to be producing 100,000 batteries a year in 2024 for Juke, Qashqai and other models, too.

Family history also made great strides, after taking out an

The Lee Adams Story

online subscription to *Find my Past* from where details of births, marriages and deaths and census returns could be easily accessed. As a result, I decided to write up the findings as *The Lee Adams Story* and completed it in January. That described my paternal family who, like many others worked on the mantra of 'Do not ask; do

134

not tell'. Soon I repeated the exercise with regards my maternal, Thomas, family during research in February and March and printing in April. As the tinplate industry featured so strongly in its lives, the title chosen was *Tinplate Tales*. Although it was finished in April, that was partly because the difficulties, of identifying precisely who the owners were of such a commonplace surname of Thomas, resulted in an early decision to terminate the work.

Just before *Tinplate Tales* was ready, early In March, Sylvia's sister, Joyce, died. A month later Margaret Thatcher also died. Two years later Joyce's husband, Derek Ridings, followed.

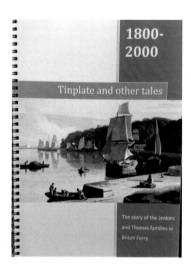

Tinplate Tales

Chrysler 300

2014

There was sufficient local, as well as family, history in *The Lee Adams Story and Tinplate Tales* that I had to give serious thought to incorporating much of the material from both books as the basis for an updated, and published, history of the town in which the Adams and Thomas families lived. Fortunately, John Fleming of Internet TSP was at hand to advise on book design, production and publishing matters as well as web design and payment systems. He operated, most conveniently, from further down Corve Street. The result was my first, and first published, book entitled: *A Most Industrious Town: Briton Ferry and its People: 1814–2014* and Briton Ferry Books was now up-and-running as a publisher.

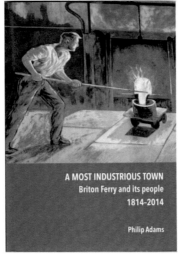

A Most Industrious Town: Briton Ferry and its People: 1814–2014

Late in 2014, or maybe early in 2015, new neighbours had moved in after the Collinses had sold 143 Corve Street. We had not yet met them and were uncertain of their name. One day, I was in the garden and saw my neighbour for the first time, as he was walking towards me to reach his back door. Even before we exchanged pleasantries in passing one another, he bore a resemblance to someone I had known … some fifty years previously.

Afterwards I began to ponder … could it have been? Maybe we should have introduced ourselves. Or did he just give his name as Jack and me give mine as Philip? Did he not have a trace of a south African accent? And the name Jack? It must be! A few days later I put the question to his wife Sue: is your surname Spence? Indeed, it was. It was almost fifty years since I last saw Jack at the University in Swansea. Then he was Lecturer in International Relations when I was taking a related paper on *Soldiers and Government*.

So, where on earth had he been during the intervening time? From 1973 he was appointed Professor of Politics at the University of Leicester and pro-vice Chancellor before becoming Director of Studies at the Royal Institute of International Affairs until 1997. Finally, prior to 'retirement' he was appointed Professor of War Studies at King's College, London in 1997, picking up an OBE in 2002 for his work for the MOD. And he's still writing, but can't say about what because I, too, am sworn to secrecy.

Jack was always very proud of the achievements of his students from the south Wales Grammar Schools. In one discussion, he mentioned Philip Williams (always Phil) once of Cwm and Neath. He had been a friend of mine and was in touch again after he had come across my books. To cut a long story short, I was pleased to arrange for Phil to come to Ludlow on his next visit to Neath. Here is his current story, as told on the University of Pittsburgh website:

> As the Posvar Chair for International Security Studies and Director of the Matthew B. Ridgway Center, (Phil) conducts several students working groups, including the Russian Contract Killer Database which has compiled extensive evidence on Russian contract killings since the 1990s. Aside from his extensive consulting work for organizations like the United Nations and the CIA, he has also given Congressional testimony on organised crime.

2015

Some interesting information in *A Most Industrious Town*, which was particularly devoted to the town's industrial, political and social history deserved augmenting. With further research into the First World War in Briton Ferry, it became clear that there was significant opposition. This resulted in a second book called *Not in Our Name: War Dissent in a Welsh Town* which detailed the nature of that opposition. Three publicity presentations for it were made at Briton Ferry Library, one in Neath and two more in Swansea.

Family interest in that opposition arose because of two of the town's conscientious objectors. One, John Adams, was my uncle. Another, Tom Thomas, was my grandmother's eldest brother. I knew almost nothing about them before undertaking this work but had long suspected there was much to tell. The town was very unusual in attracting many very distinguished figures, from Wales itself but also far beyond, so that its history possessed far more than simply local interest.

Experts who specialised in the opposition to conscription and war during World War One, such as

Not in Our Name detailed the nature of the opposition to war.

Cyril Pearce of Leeds University, advocated that local, not national, studies were required to understand the issues involved. *Not in Our Name* started to fill that void for Briton Ferry, closely followed by another book on Port Talbot.

My last paid work was at Duresta, a furniture manufacturer at Long Eaton, Nottingham, for Tern Associates. In September, I attended my first meeting of Ludlow Sustainable Transport Group.

In Ludlow, an *In and out of Ludlow* group had been set up. Its main aims still are to provide a forum for individual residents, councillors, and organisations to seek consensus on traffic, transport, parking and access issues in and around Ludlow, and to improve the experience of residents and visitors travelling to, around and from the town.

In September, as a member of Ludlow 21, I joined the first meeting of Ludlow Sustainable Transport Group. Its aim is to enable the *sustainable* movement of people and goods in the area, in terms of reducing carbon emissions and achieving health benefits, by applying the principles of active travel and the concept of avoiding, shifting, and improving the transport modes that people use. The group considers pedestrians, cyclists, buses, trains, and motor vehicles.

Richard Olsen and I were initially tasked to survey Ludlow businesses regarding the provision of electric vehicle charging.

2016

Daring to Defy: Port Talbot's War Resistance 1914–1918 was the next book, about the neighbouring town of Port Talbot's experiences regarding opposition to conscription and war during World War One. The content was again factual, granular, and unchallengeable. Together, the two books created a cumulative impact which led to Cyril Pearce's comment, in 'Communities of Resistance', his go-to source for everything you want to know about the peace movement and conscientious objection in the first World War:

> *The anti-war communities of Aberavon and Briton Ferry were unique. There were no other places quite like them. They were probably the most significant and effective in any area of Wales or any other part of Britain for that matter.*

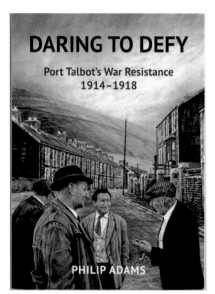

Daring to Defy:
Port Talbot's War
Resistance
1914–1918

There was resistance, too, in Port Talbot to plans announced in March by Tata Steel to sell its loss-making UK steel business, putting at least 15,000 jobs at risk at sites including Port Talbot, Llanwern, Rotherham, Shotton, Corby, Redcar and Hartlepool.

On 23 June 2016, the UK held a referendum on its membership of the EU. The question facing voters was: 'Should the United Kingdom remain a member of the European Union or leave the European Union?' 51.89% of voters voted to leave the EU. The UK left on 31 January 2020. Nigel Farage claimed in 2016 that voting to leave the EU was the UK's only chance to save its steel industry.

Neath-Port Talbot, which had been predicted to be one of the fifteen closest results in the referendum, voted strongly to leave the bloc. The Press Association reported 57% of voters in Neath Port Talbot backed leaving. That compared to a prediction of 50.5% for remain and 49.5% for leave in analysis published by J. P. Morgan ahead of the poll.

In Sunderland, the turnout was 64.8%, with 61.3% voting leave,

The Port Talbot Tata Steel plant was under threat of closure after the company announced plans to quit its entire British steel operation earlier in the year, putting thousands of jobs at risk. The anti-EU UK Independence Party had seized on the potential collapse of the plant, saying the EU did not do enough to stop cheap Chinese imports and the bloc's rules on state aid have prevented government intervention.

The hubbub that enveloped the Brexit debate and Tata's disquieting announcement meant that the two-month long Fort McMurray fire, in Alberta, Canada, received scant attention in parts of Britain. Seven years later, its significance for the need of decarbonising, would be all-too apparent at Port Talbot, Scunthorpe Grangemouth and elsewhere.

Not all were breaking away from one another: in the world automotive industry a Nissan-Renault-Mitsubishi Alliance was inaugurated in 2016.

2017

One of our frequent visits to Italy which, this time had some unexpected consequences, involved Catania in Sicily in February. Absolutely unaware of what the Festival of St Agathe involved, our train trip around Mount Etna took place in the most surprising context of massive street celebrations ... and packed hotels!

Yorkshire featured several times with a reunion of Briton Ferry schoolmates in Bradford, an invitation to the Ebor Races in Doncaster from niece Rebecca Adams and a 70th birthday at York for her father, Andrew.

'Duffy-Ferraro' family Christmas event at Tenbury Wells

On industrial matters a visit to the former pit village of Llay, near Wrexham, to attend a Welsh People's History presentation at the handsome Llay Miners' Welfare Hall to hear of the Llay experience was most rewarding both in terms of the history of the pit and its welfare function of 'providing a social welfare centre and recreation grounds for the use of its beneficiaries with the object of improving the conditions of life for the beneficiaries ... via the provision of sporting, recreational and social facilities'.

Although no book arose from it, I prepared yet a further study of opposition to conscription and war in the Swansea area. Llafur, the Journal of Welsh People's History, published it as *Conscientious Objection in the Swansea area: Cleaver re-visited*. Together with the previous books, this just about completed the picture of opposition to war and conscription in the steel, coal, and tinplate towns of west Glamorgan. A paper on an individual objector was entitled: *John Adams: A story of conscientious Objection* was published in Volume Three of *The Neath Antiquarian* in 2018.

When Martin Davies and I were researching the railways of Briton Ferry and district, a most bizarre event occurred on our only visit to Briton Ferry station. Whilst interviewing departing and arriving passengers, one approached me to say: 'Do you know who I am?' It took me some time to figure it out, but he turned out to be former schoolmate Michael Hughes whom I had likely not seen for 50 years. His train journey would end in New Zealand where he would settle with his children, his wife having recently died. The odds against us being there at the time of his departure were immense. Michael was not one of the 'Ferryboyz', the name of the group of schoolfriends who had reunited from time to time.

2018

Gareth Rees, a member of the Ferryboyz, persuaded our group to visit Hamburg, with the main attraction being a visit to St Pauli FC. The other was to see Hannover v Bayer Leverkausen. The difference in atmosphere between the two games was remarkable. At fan-owned St Pauli the atmosphere was electric, particularly when the home side came from behind right at the end to beat Holstein Kiel. In contrast the Hannover match totally lacked ambience.

The importance of railways in the growth and geography of the coal, steel and tinplate industries in the area had barely been covered in detail in any of the previous books. A friend of longstanding from Briton Ferry, Martin B Davies, was keen to do so. We agreed to co-author and publish a book entitled *Reshaping rail in South Wales: the railways of Briton Ferry and district-past, present, and future*. He wrote most of the book, concerning historical aspects, whilst I considered the present and future. Martin made publicity presentations at Briton Ferry, Neath and Pontardawe, and I followed these up in Port Talbot at the invitation of the south Wales branch of the RCTS (Railway Correspondence and Travel Society). I urged Martin to always insert the B between Martin and Davies to distinguish himself from many other Martin Daviesies.

Railways of Briton Ferry cover

Peace Train logo

That autumn, my interest and knowledge of railways was called upon by Kim Holroyd to organise the logistics for a *Peace Train* on United Nations Day. The objective of the plan was to take a petition from Hope, near Flint, to the House of Commons in support of the International Campaign to Abolish Nuclear Weapons (ICAN). Dr Rebecca Johnson, Nuclear Physicist, Nobel Peace prize winner and founder member of ICAN, welcomed the strong support at all stations on the three rail lines involved. Chester, Wrexham, Ruabon, Chirk, Gobowen, Shrewsbury, Church Stretton, Craven Arms, Ludlow, Leominster, Hereford, Abergavenny, Cwmbran, Newport, and Swindon stations offered mail bags containing thousands of letters to Ministers and MPs, to be sorted on board for delivery to MPs at the House of Commons.

St Pancras kiss

That autumn, Ludlow French Twinning started a French film season, at Sylvia's suggestion. Held monthly at the Blue Boar, it has continued to run successfully as an event for members and friends and raise funds for subsidising children's travel.

A reunion for former Amicus tutors at Esher was just that, with no proposals for further work, but it was a last reunion because we never got together again.

Sylvia and I paid our 55th trip to Italy since the Millenium, but this had become a visit of diminishing frequency as each of the four Ferraro grandchildren left home in Poggio Mirteto to study at Scottish and Irish Universities: Glasgow, Stirling, and UC Dublin in that order, so far.

As part of a bigger project to make 144 Corve Street greener and more energy efficient, I both joined Ludlow car club and bought an electric car, installed solar panels on the garage roof and put battery storage in the garage. The Nissan Leaf car was not the first electric vehicle (EV) in Ludlow, because someone already had a Tesla model. Afterwards I was interviewed on Shropshire radio regarding EVs to speak on behalf of Ludlow Town Council. I explained the council's plans for on-street EV charging and the part played by Ludlow 21's Sustainable Transport Group in informing the Council.

Even electric SUVs are heavier and bigger and not friendly

Riversimple's HYRBAN demonstrator. The company pioneered zero-emission hydrogen cars in Ludlow.

2019

A post-Brexit *People's Vote* was still being called for and Sylvia and I attended both local and national events in support of it, with Sylvia organising several. A year of much European travel took us, coincidentally, to many ports. Ireland (Waterford) and Scotland (Fort William to Mallaig by steam train), France and Italy (Bordeaux, Genoa, and Marseilles). Travel was not easy with a French rail strike over Christmas, but one unexpected bonus of the strike was to see a son-et lumière at Amiens Cathedral to commemorate its 800th anniversary during the Christmas market period. The visit was brought about by the need to pay an urgent evening visit to the nearby SNCF ticket office to check whether our planned train was running the following day. Some trains ran, and one got us to the Belgian National Rail Museum at Schaerbeek in Brussels.

Amiens cathedral's 800th anniversary

A trip on *The Jacobite* over the west Highland line from Fort William to Mallaig nearly did not happen. That was because I mistakenly binned the Christmas gift voucher for the trip, thinking it was just promotional material. West Coast Railways honoured the booking, nevertheless.

Less than a year afterwards the Covid epidemic would follow. This would mean that the pleasant memories of such events as the International Meccano Exhibition at La Ferté-Macé and the delightful visit to Fougères would just stay in our memories for a little longer. The *Ferryboyz*' visit to Portmeirion was completed just in time.

Covid walks enabled one to visit places such as Pilleth

Montgomery Canal

2020

In January, when Coronavirus was reported in Wuhan, I was creating additional Wikipedia pages on *Briton Ferry*. With great support from Ludlow 21 members who distributed my election leaflets, I stood for election to Ludlow Town Council in February saying that:
> My aim is to anticipate issues wherever possible before they become problems for residents, and to deal with them positively. Where a resident in the ward raises any matter with me, I will investigate it impartially to achieve the best outcome, but I will not make promises that I cannot keep.

Just before lockdown, Ludlow Sustainable Transport Group presented a proposal to Shropshire Council's portfolio holder on a scheme for electrifying Ludlow Town bus services which Councillor Jim Smithers and I had prepared. Council took an interest in the scheme, agreeing that it was self-financing if grants to purchases the buses were forthcoming. Despite many promises from central government nothing has been forthcoming.

I considered that I had become a 'Covid Councillor'. For quite a period afterward proceedings were held on Zoom and councils operated under delegated powers, making the role of a councillor somewhat different to normal. In March, after the first UK death from Covid, a pandemic was declared, and everything changed.

A group of volunteers called *Pulling Together Ludlow* was formed as an emergency initiative of Station Drive and Portcullis surgeries. It was in collaboration with *Hands Together Ludlow* to support the residents of Ludlow and nearby who needed social and non-medical attention during the Covid crisis. This included shopping and prescription deliveries, meal deliveries for those eligible, a telephone helpline and an offer of social telephone calls.

Nick Young chaired the group as overall co-ordinator and MP and NHS liaison; I was Secretary; John Wallwork provided volunteer support; Chris Deaves managed IT and telephony systems, Isabel Barber Chavez ran medical liaison and volunteer registration, Victoria Harris controlled Facebook communication, Pete Gray co-ordinated volunteers, with Erica Garner liaising with the voluntary sector, community, and local government.

Pulling Together Ludlow met online daily by Zoom for the first time on 19 March and for the last time at meeting 47 on Monday 27 July. Shops had re-opened in June. Thereafter, all support operations were to be provided, as normally, by *Hands Together Ludlow*. Should the Covid-19 emergency demand it afterward, *Pulling Together Ludlow* would return to operating on an emergency basis.

At one point during the crisis, it came to our notice that 16 Via Marsala, the house that Sylvia had bought on Poggio Mirteto to visit as the grandchildren grew up, had been flooded from the apartment above, rendering it uninhabitable.

Covid enabled me to resurrect my railway writing by contributing articles to *News from the Marches Line*. I contributed to six editions until August 2022. The titles were:
- *A trip to Train World Brussels*
- *From Brussels to the Gers*
- *Hadrian's Wall Country Line*
- *From Red to Green-rail and the environment*
- *Linguistics along the line: Ludlow to Dundee*
- *The Volcano Train*

2021

On 26 January the death of the 100,000th Covid victim was announced. Living in rural Shropshire we were very fortunate to be able do interesting country walks in south Shropshire and north Herefordshire, based on books illustrating interesting features which appeared on the walks. Two favourites for walks or bike rides were Chirk and Montgomery canals.

Chirk Canal

At our age we also received Covid vaccinations, the first at the end of January and a second in April. Home testing became commonplace but overseas travel required special documents like the Passenger Locator Form and evidence of vaccination until March 2022. In June outdoor markets and non-essential retail were allowed to open.

North Shropshire MP Owen Patterson's resignation came in November after a report by the Parliamentary Commissioner for Standards found that he had broken paid advocacy rules. He lobbied for Randox, a firm that provided the Department of Health and Social Care with test kits which were required to be used to complete such documents as the Passenger Locator Form. Randox cost us money which was not repaid.

On 22 May, south Shropshire Climate Action convened a meeting to consider its plan via Zoom for Ludlow Parliamentary Constituency to reach net zero carbon emissions by 2030. It comprised the four action areas of:
- Land management.
- Energy and buildings
- Transport
- Communities and education

Each area comprised a team and I joined the transport team along with others from LSTG and elsewhere. The team was subdivided into four areas. I was assigned the topic of car sharing and its role in reducing CO_2 emissions in south Shropshire. I became increasingly disappointed with the way the group was run. Working in isolation, which I thought was a poor idea, I got through a lot of work. Little of what I suggested in support of the concept of *'Why drive around sitting next to an empty armchair with an empty sofa behind you?'* was used. In the real world, members of Ludlow U3A embraced car sharing for members when hosting at their homes, in an area from Greete to Wigmore to Lydbury North.

From June, I chaired the Town Council's Policy and Finance Committee. I considered that financial aspects were being managed well but that policies were not. The two major deficiencies with policy were that scrutiny was lacking and that access to policies and procedures was often difficult for Councillors. Accordingly, I proposed to Council that:
- a universal template be adopted for all policies, procedures, and administrative processes to assist in their preparation and presentation.
- These be held in an electronic library, each with a unique reference number.
- They are organised to reflect the council's executive and service functions.
- The Policy and Finance committee determines who has access to what.

- The existing practice of reviewing three policies per meeting be retained, but with proper scrutiny and no duplication.

In terms of making most of things, when attending the funeral of Sylvia's cousin, Shirley, we took advantage of its proximity and paid a pleasantly tranquil visit to Bletchley Park Museum to 'discover incredible achievements of Britain's World War Two Codebreakers, in the place where it happened'.

Bletchley Park main office

Unknown to her, I had been making great plans for Sylvia's 80th birthday in August. Initially the surprise reception was to take place at the Station House in Acton followed by visits to Kew Gardens and her birthplace in Ockley. The London-based Adams 'children' were most helpful in setting up the Station House event for friends who had travelled from different parts of the country. Some did not make it because of work (Rebecca Adams was in Japan) or Covid (Daisy Ferraro had it, as well as a fractured fifth metatarsal). Happily, old friends from Durham made it, to Sylvia's delight and surprise but daughters Ruth Ferraro and Jenny Dewar were prevented from travelling from Italy and France.

On 30 July, Honda Swindon closed after thirty-six years of operation to end its production of cars in Europe. In its

153

time it had turned out the Accord, the Jazz, the CRV and its flagship car, the Civic. 3,400 staff lost their jobs. A trade deal in 2017 between Japan and the EU had put an end to tariffs on car imports between the countries involved. 'Brexit uncertainties' were also cited by Honda as a reason for the closure. Another reason given was the need to change over to the production of hybrid and electric vehicles. One questions why the trade deal did not result in similar action by Nissan and Toyota.

A Christmas visit to France was cancelled, but on a beautiful bright day, Councillor Adams instead rejoined Pulling Together and Council colleagues to deliver Christmas gifts around Ludlow. It was the first Christmas at home for quite some time – and all because of Covid!

A surprise guest at Kew

2022

After the Russian invasion of Ukraine in February, the Ludlow area hosted several Ukrainian families. We did not host young Artem and Kyril but looked after them for the odd day during the school holidays when their mother worked in the nearby Feathers Hotel. They soon learned English. Sylvia used her experience of intercultural competence to introduce aspects of British life to curious Ukrainians young or old. I learned much, too, including from the two boys who told me about their home city and showed me pictures of their bombed school in Mykolaiv, a port city on the Black Sea of almost half a million population.

The battle of Mykolaiv had started on the night of 26 February 2022. Russian forces were repulsed from the city in March, and by April all but a few of its surrounding villages were back under Ukrainian control. On 4 March, it was seen as the next key stepping-stone for Russian forces on the road to Odessa.

Artem and Kyril take off their Halloween masks

Bonjour was the only complete record of any UK Twinning Association

The cancellation of the previous year's twinning visit to France, due to Covid 19, prompted me, after fifteen years membership of Ludlow French Twinning, to account for its 36-year history in a fifth book called *Bonjour: Ludlow French Twinning – a unique experience*. It would become the only complete record of any UK Twinning Association, conceived during a period of Covid isolation in May and delivered from the printer in September. Once again, I published it through Briton Ferry Books. Had the pandemic continued, would twinning have survived as it did after Brexit? Physical memorabilia of twinning existed but what about people's memories of so many instructive and enjoyable events here and in France?

Even having prepared almost fifty editions of Ludlow French Twinning's Newsletter, which provided much for the book, they only covered a decade of the Association's twinning activity. The rest came from minutes of meetings and, more importantly, from the mouths of founding members and those whose subsequent enthusiasm, ideas and commitment kept things going.

It explored the factors needed for successful twinning, such as relationships with local government, and the influence of Ludlow twinning's food and drink, sporting, and musical connections.

It was, above all, about the people who have made

twinning possible and those who have benefitted from its unique experience over the years. Interviews with the likes of John and Janice Holliday, George and Lauren Hough, Linda Senior and Dave Mulliner captured their personal experiences which demonstrated the enjoyment and many mutual benefits of twinning.

Between April 2022 and March 2023, Trussel Trust food banks provided almost 3 million food supplies to people in crisis, a 37% increase on the previous year. Today there are 2,600 food banks in the UK.

In early July, 62 of the United Kingdom's 179 government ministers, parliamentary private secretaries, trade envoys, and party vice-chairmen resigned from their positions in the second administration formed by Boris Johnson as Prime Minister, culminating in Johnson's resignation on 7 July. He continued as caretaker until 5 September when he was succeeded by Liz Truss. On 20 October she resigned amid another government crisis, making her the shortest-serving prime minister in British history.

Rachel Adams and Edd Setterington were married at Marylebone Town Hall just before Christmas and we attended en route to Italy. There were train strikes at that time and it was quite an achievement to get to both the wedding and Italy and back. A bonus was our visit to the BBC studios at Portland Place on 18 December, courtesy of niece and sports journalist Rebecca Adams. We saw parts of the France v Argentina World Cup Final from Lusail, Qatar, but our real interest was in seeing the studios where she often worked.

In September Queen Elizabeth II died. Our neighbour, Nick James, was taken ill after a fishing visit to Scotland and after much time in hospital also died in the New Year, with symptoms of Covid being involved in his death.

2023

It was not just the present aggression in Ukraine that was being reported. Past aggression was referred to in a somewhat more cheerful context. In April, sister-in-law Mary Adams, attended her father's 100th birthday at his home near Trim, Co Meath. Barney O'Dowd's birthday was 'a remarkable milestone for anyone', said Martin Doyle of *The Irish Times,* 'in that he almost died 47 years ago'. In January 1976, he was shot and gravely wounded by loyalist paramilitaries in his farmhouse in Ballydougan, Co Down, during a family gathering at the end of the Christmas holidays. His eldest brother Joe (61) and two of his sons, Barry (24) and Declan (19), were murdered in the attack. Among the dozens of cards Mr O'Dowd received to mark his birthday was a letter from President Michael D Higgins, accompanying his centenarian bounty of €2,540. In a handwritten postscript, Mr Higgins wrote:

> *I am conscious, I assure you, of the great sorrow inflicted on you, and personal suffering too, which makes it even more important, even at a distance. I will salute your indomitable spirit on the day and recall those taken from you.*

Barney died on 9 April 2024, the very day the final draft of this work was sent for typesetting.

Despite the train strikes, the 'Ferryboyz' organised a further reunion, this time in Glasgow. Instead of watching St Pauli FC in Hamburg against Kiel, as we had done in 2018, this time it was Rangers v Aberdeen at Hampden Park. The train ride to and from Glasgow Central to St Florida for the afternoon match was interesting, but the journey of most interest was the twenty-two stop, open-top, all day, hop-on and off bus ride from George Square

A long overdue civil partnership

to places like Glasgow Green, the Riverside Developments and Kelvingrove Park and Gallery. Later in the year Sylvia and I repeated the idea in Turin, where the system comprises four coloured tourist bus routes which visit different parts of the pleasant, de-industrialised city. They include the historic Fiat factory with the test track on the roof and the Juventus FC Stadium in the Vallette borough of the city.

After twenty years together, Sylvia and I entered a long overdue civil partnership in February in Shrewsbury. We had hoped it would take place the previous Autumn, but Covid had delayed things. In fact, we were still receiving vaccinations at places as distant as Pontesbury at this time. However, the pre- and post-ceremony partnership events were a family affair in Ludlow Castle and Fishmore Hall; the weather was generally kind, and everything went well.

In June, BBC TV transmitted a documentary series called the *Steeltown Murders* about three local murders in 1973. It examined how the identification of the perpetrator, Joseph Kappen of Port Talbot, was finally made through familial DNA profiling and his body's exhumation in 2003. This brought Port Talbot to national attention before it was again in nationwide focus because of steelworks job losses fewer than three months later. An opportunity arose to see Neath Abbey and its nearby

A long overdue civil partnership

159

Inside Neath Abbey

ironworks during an open day whilst visiting the Davieses in Pontardawe. Despite living less than three miles away during my childhood, I was ashamed to say that I had visited neither, despite having been to similar sites at Coalbrookdale and Blaenavon several times.

The Cistercian abbey, Wales' largest, had long been surrounded by industry. The ironworks is a site of world importance, producing iron in its two 18th-century blast furnaces. At the same time, it has been an historical engineering location, at different times making railway locomotives, marine engines, iron ships and stationary steam engines.

Behind the Ironworks is a viaduct designed by Isambard Kingdom Brunel which carries the mainline from London right over the site. The engine manufactory was one of the earliest buildings to have cast-iron roof beams – and a row of workers' cottages.

The Fox family, who were Cornish Quakers, built the ironworks one sees today in 1792, to supply their own foundry and the open market. They were great exponents

Neath Abbey furnaces

of vertical integration, whereby you own all the businesses in your supply chain – which in their case meant owning everything from the mines to the end products made with the iron: the engines and ships. This business model made them hugely successful and brought Neath Abbey Ironworks worldwide renown.

Joseph Tregelles Price, the company's managing director from 1818, was a famous philanthropist. He had co-founded the Peace Society in 1816 and tried to win a reprieve for Dic Penderyn, arrested in the Merthyr Rising and hanged in Cardiff in 1831. Price also campaigned for the abolition of slavery and set up a fund to buy a Bible for every freed slave.

Ludlow Rail Users Group, which had been one of the Marches line users' groups, was dissolved, despite hopes that it might be rejuvenated in conjunction with Ludlow 21 to explain and encourage the environmental benefits of train travel over other modes of transport.

The disruption on the railways due to industrial action also applied to ferry services: Ludlow French Twinning had to cross the channel from Newhaven to Dieppe instead of Portsmouth to Caen, but the rerouting provided travellers with the chance to spend an informative day at Rouen. Most interest was in the Place du Vieux-Marché, where Joan of Arc was burned at the stake for heresy in 1431 and where the small garden, *Le Bouchet*, which is outside the church marks the exact spot.

On 5 June I resigned from the Town Council because of my concern about the agenda-setting process for the Policy and Finance Committee. I had received the Town Clerk's assurance, in the form of a draft agenda, that an item was listed 'to review and approve the Environmental Policy Statement.' I informed the Chair of the Environment sub-committee that was the case. That item had been removed when the official agenda was issued. Once the item was put there it should not have been removed, but any concerns about it should have been explained and discussed at the meeting.

That was my immediate reason for resigning. I left with mixed feelings because I thought that policymaking was progressing quite well. Soon, however, would come the time where it would be necessary to computerise the policy library and I had my doubts whether there was sufficient enthusiasm to do so. In the background, however, more fundamental matters in the council needed attention. Most modern organisations encourage teamwork and continuous improvement on top of basic practices like induction training and daily staff briefings. I was not aware of any.

The culture seemed to be based on claims that the Council was insufficiently staffed. Whenever a change was advocated, the possibility was stifled, on the grounds that the resources to do so were said to be unavailable. My view was that better management would free up any resources needed. On top of this was the high turnover of councillors and the fact that available expertise was not called upon. The council, therefore, had no collective 'memory'. It was time to go.

Later, in the Autumn, I was privileged to make two further book-related presentations. The first was entitled *Crossing Borders* in the delightful garden of Ludlow's Castle Bookshop. This was a bold attempt to describe the evolution of my writing and to trace any thread that ran through my books. The second was for Neath Antiquarian Society. Entitled *Opposition to War in west Glamorgan*, it showed how the industrialisation of the area had led to a rich local culture and much municipal enterprise and radicalism. Some of that was directed into opposition to war and conscription and to resisting the demands of the state, whether for religious or political reasons.

It was also time for newly-married Rachel (née Adams) and husband Edd Setterington to go to the US to work, so we said goodbye to them in Northumberland in July.

In September, Tata Steel announced that it is to invest £1.25 billion, including a UK Government grant of £500 to finance a decarbonisation project to enable greener steel to be made at Port Talbot. The steel would be greener because it would be made in electric arc furnaces which emit four times less CO_2 than blast furnaces The project entailed winding down blast furnaces and coke ovens at a cost of 2,500 job losses within 18 months and 300 more to follow. The announcement was made when Tata claimed to be losing £1 million a day at the plant due to energy costs and other factors. UK energy costs 1.5 to 1.6 times more than in France and Germany. The *Daily Telegraph* newspaper called Port Talbot 'a victim of Net Zero madness', but it was not madness at all, just a necessary

163

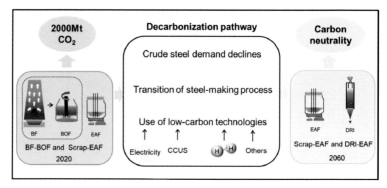

Decarbonisation pathway for steelmaking (Tata Steel UK)

step to cut CO_2 emissions, avoid climate breakdown and cut such losses.

Many considered that it was not the decision to decarbonise that was wrong but the phasing of its implementation. Its abruptness was reminiscent of the closure of the coal mines in the 1980s and the lack of sufficient consultation and preparation for replacement employment. Two Union groups each presented alternatives to the HMG/Tata announced plan. They were the Unite *Workers Plan* and the GMB/Community's *Syndex* plan.

HMG/Tata created a Transition Board to develop skills and regeneration as part of the project, with working groups to deal with outplacement, supply chain matters, community relations and communication and wellbeing. In October it first met and recognised 'the urgency with which we need to work'.

2024

For Welsh rugby, it was the year of the Wooden Spoon for the first time since 2003. When visitors Italy beat Wales 21 to 24 in Cardiff, they had recently beaten Scotland and drawn with France. Italian prowess with the oval ball had been developing since the team had joined the International Rugby board in 1987. Many in Port Talbot attributed some of the Italian success to the foundations laid a half century earlier by the country's first National Coach, Port Talbot's Roy Bish of Aberavon RFC.

The televised screening of *The Way* in February was eerily applicable to actual events about to happen, that were happening at the time and had previously happened. Michael Sheen's direction of the series concerned civil unrest in Port Talbot when the government hired private security contractors to protect agency workers at the steelworks. It followed the death of a youngster in molten slag in the steelworks and his father's suicide that followed. Something similar had happened in 2006, but without civil unrest. The unease about plant closure announcements which had been made the previous Autumn reinforced the series' relevance.

Unease, too, was expressed about the removal from the town of street artist Banksy's mural *Christmas Greetings*. Yet, as with the 2006 tragedy, the town reacted positively to its loss. The recovery takes the form of an *Urban Street Hub* which now features over sixty large-scale symbolic murals featuring the area's people, places, and events.

It is worth comparing the situation elsewhere world to see how other industrial organisations are dealing with the need for the decarbonisation arising from climate change.

Danieli electric arc furnace

In Britain, British Steel, under the ownership of Jingye Group since 2020, has planning permission to build the first of two electric arc furnaces at Lackenby, Teeside, with the second planned for Scunthorpe as blast furnace replacements. The Department of Business and Trade is supporting with £300 million funding within British Steel's £1.25 billion decarbonisation plan. The plants will continue to produce construction steel, rail, wire rod and intermediate products.

In the auto industry, Chinese firms like BYD and Li Auto are fiercely competitive and can sell their electric cars in the UK 20% cheaper than its European competitors, thus gaining market share. As a result, Nissan and Honda, for example, will co-operate on the development of EV components and software. Plants will adopt Nissan's *Intelligent Factory* concept to reduce production time by 20% and production costs by even more to compete with models like MG, produced by Chinese state-owned automaker SAIC Motor Corporation. The Intelligent Factory initiative at the Nissan Tochigi Plant, will be introduced at Sunderland to 'produce vehicles for a decarbonised society' by:
- Using robots that have inherited the skills of monozukuri (manufacturing staff) and takumi (master technicians) to make next-generation vehicles of the highest quality.

- Creating an improved environment where a wide range of people can work comfortably, and
- Realising a zero-emission production system, contributing to a decarbonised society. All electricity used will be generated from renewable energy sources and or generated with onsite fuel cells that use alternative fuels.

Other recent overseas investments in the UK car industry are Tata's Jaguar Land Rover's £4 billion battery plant at Bridgwater, BMW's Mini plants at Oxford and Swindon. Stellantis at Ellesmere Port which is Britain's only EV-only plant will make vans. Envision AEC/Nissan will build a second battery plant at Sunderland.

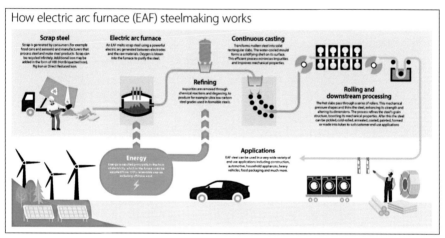

How EAFs work
(Tata Steel UK)

Decarbonising or deindustrialisation? Port Talbot 2024

In September 2023 Tata Steel announced a plan to end steelmaking at Port Talbot via the blast furnace-basic oxygen route in favour of the electric steelmaking route. The changes, which would entail the loss of some 2,500 jobs were described as 'decarbonisation', and 'green steelmaking'. They were also made against claimed daily losses of £1million.

Tata supplies around a million tonnes of steel to meet 50% of UK carmakers' requirements. This represents 35% of sales (by revenue) for body panels, chassis components and wheels to manufacturers like BMW, Nissan, and Jaguar Land Rover in the UK, as well as to customers in mainland Europe and elsewhere. This is predicted to increase jobs in coming years as carmakers move to electric vehicles following a downturn in sales during the coronavirus pandemic.

A group known as the Trade Union Steel Committee presented alternatives to the HM Government/Tata announced plan. Tata rejected most of the plan because it was unable to agree to the proposal to retain blast furnaces, for both financial and technical reasons. In November 2023 the Unions' alternative plan became two in the form of the Unite Workers Plan and the GMB/Community's Syndex plan. The Unions claimed that reducing UK energy costs, which are 1.5 to 1.6 times more than in France and Germany and a requirement for HMG to procure UK made steel in government contracts might attenuate the £1 million a day loss.

Much of Tata's plan has come through HMG press releases. Detail regarding the precise timing of events, such as the closure of existing plant and the capabilities of the replacement technology has been vague. However,

HMG/Tata announced in February 2024 the creation of a Transition Board to develop skills and regeneration, with working groups to deal with outplacement, supply chain matters, community relations and communication and wellbeing. HMG contributed £80 million and Tata £20 million to finance the Board's work. The board first met in October 2023 and 'recognises the urgency with which we need to work.'

The coke ovens and No 5 blast furnace were scheduled to be shut down in June 2024, but at the end of March it was announced that the coke ovens would close immediately due to 'stability concerns', with coke imported from elsewhere. Trade unions suggested that such decisions should be left until a general election was held, which would be no later than October, in the expectation of a Labour Government being elected which would be more in favour of its plans.

In the event 4 July was announced on 22 May as the date of the General Election. On the 30th the Leader of the Opposition, Keir Starmer, urged the PM to 'talk with those workers'.

'I've been there: I've looked them in the eyes, and I have told those workers I will fight for every single job they have there and the future of steel in Wales.' The results of the Sixth meeting of the Transition Board, scheduled for the end of May, took place on 11 July under Labour.

Elsewhere in the world other industrial organisations are also dealing with decarbonisation arising from climate change.

China makes 1018 million tonnes of steel, 54% of the world's output. With Japan, which makes 4.7%, these are the world's biggest steel exporters. In comparison, the UK made 6 million tonnes in 2022 but imports 61% of its steel at a value of £2 billion annually. It is significant that in 2021 alone, China replaced almost 30 million tonnes of blast furnace capacity with electric furnaces (EAF's) in following its decarbonisation pathway towards CO_2 neutrality.

In Japan, steel companies have also been abandoning

Steel imports at Briton Ferry wharf

blast furnaces for some time in favour of EAFs. For example, JFE Steel, one of Japan's 'big three' producers, is replacing its no 2 blast furnace at Kurashiki with an electric furnace. It has a tap weight of 420 tonnes and a tap-to-tap time of fifty minutes. The company has successfully produced bar, beam, hot rolled coil and high tensile auto steels using the EAF process. The common factor, therefore, within the steel and other industries worldwide is the need for a decarbonised economy to push back climate change.

In publicising its aims, the Transition Board has referred to other initiatives in south Wales which should be regarded as complementary to the steel plan: the Celtic Freeport and the Swansea Bay City Deal. (The author considers the west Wales Metro scheme to also be complementary). The timely disclosure of Tata's plans to all stakeholders will be crucial, so that they can be properly scrutinised with alternatives considered. Then it will be the effectiveness with which the Transition Board

Scrap exports at Briton Ferry wharf

and its two working groups interact with the other initiatives that will determine how smooth the transition to decarbonised steelmaking will be. The groups' roles are now described:

People, Skills and Business working group
The group is tasked with *immediate* response and support to:
- Undertake appropriate gap analysis (e.g., available jobs; training placement shortages etc) and to use data to make recommendations.
- Ensure the right support at the right time for employees directly affected, including the provision of re-employment, re-training, welfare advice, pension advice etc.
- Support those wishing to set up their own business.
- Support those businesses affected in the supply chain.
- Provide those affected with health and well-

171

being support.
- Work with local partners and higher and further education establishments to share skills programmes which are suitable for affected workers, identifying training gaps and developing new skills programmes where gaps exist.

Place and Regeneration working group

The group is tasked with the *medium to long-term* (1–10 years) response and support to:
- Work with local partners to measure the potential medium and long-term economic shock to the area including supply chain impacts.
- Work with local partners to develop an economic strategy to support Port Talbot's transition over the next decade.
- Develop and drive forward business cases to access the £100 million funding.
- Consider other support mechanisms, including wider UK Government decarbonisation measures that will facilitate the transition of the area.
- Seek and secure alternative investment that aims to at least replace the economic value of the jobs that will be lost and support replacement business opportunities.
- Establish how land holdings can be repurposed at the end of decommissioning to support the regeneration of the Tata site.
- Align and link proposals to the UK and Welsh Governments' wider strategic objectives.

Each sub-group's membership includes key partners determined locally in discussion with the Transition Board. Tables 1 and 2 below show the participant groupings and features of the four initiatives.

The Celtic Freeport

The proposed Celtic Freeport is a private-public sector partnership led by Associated British Ports (ABP), and involving Neath Port Talbot Council, Pembrokeshire

County Council, and the Port of Milford Haven. Its vision is one of the initiatives to create a green investment and innovation corridor to help attract major inward investment, skills development, and decarbonisation. The Freeport will support new manufacturing facilities and port infrastructure upgrades to service the roll-out of floating offshore wind (FLOW) in the Celtic Sea. Renewables developers, energy companies, and education providers will be encouraged to participate to provide the backbone for a cleaner future based on the hydrogen economy, sustainable fuels, carbon capture and storage, cleaner steel, and low-carbon logistics.

Swansea Bay City deal

The fifteen-year Swansea Bay City deal is an initiative which covers a larger but similar area with complementary economic, and social objectives, in nine projects at the cost of £1.3 billion is being met on a 45/55 basis between public and private investment Projects are spread along the 'internet coast' within three groupings: Internet of Energy, Internet of Economic Acceleration and Internet of Life Science and Wellbeing.

Swansea Bay and west Wales Metro

This initiative by Transport for Wales (TfW) involves studies with the Welsh Government and local authorities to develop plans to provide better infrastructure and connectivity and the potential for faster transport services between west and south Wales. It entails working with partners to change the way that transport is provided, boosting the local economy, and providing better access to job and leisure opportunities. This includes:
- faster and more frequent rail services with different route options to Swansea, Carmarthen, and Milford Haven.
- the introduction of new lines and services in the Swansea Bay Area; providing greater connectivity and new opportunities for rail travel.
- improving the speed and reliability of bus journeys; improving passenger waiting facilities,

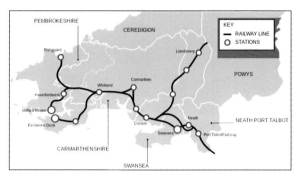

Swansea Bay and West Wales Metro (TfW)

bus lanes, and intelligent traffic signals to help reduce journey times.
* trial use of hydrogen buses in Swansea Bay and Pembrokeshire for a more sustainable approach to public transport.
* timetable integration – bus and rail timetables to work together.
* active travel – rail and bus stations to cater for pedestrian and cycle travel.
* ticket integration to include multiple journeys on one ticket, incorporating bus and rail travel, with the best possible fare option at the time of your travel.

Workers' Plan for Port Talbot

The most recent initiative is the Workers' Plan for Port Talbot, proposed by the Unite trade union. It has wider relevance than Port Talbot. The underlying philosophy is that there are two paths to the future on offer. The current one is of managed decline and that proposed is to defend existing jobs and to create new jobs within a green transition strategy. The plan's two stages are aimed at first dealing with the immediate situation, and then enhancing steel production during the second phase, within an Industrial revival for south Wales based on the decarbonisation plans shown in Tables 1 and 2. In addition to its proposals for the steelworks itself it also envisages a manufacturing centre for floating offshore wind equipment in the Celtic Sea, onshore wind including turbine manufacturing, green hydrogen and chemical feedstock production such as for green aviation fuel, tidal energy in Swansea Bay, a battery factory and solar farms and Low emission cement manufacture.

An assessment of the decarbonisation plans

All plans accept the need for decarbonisation of the industry and all parties agree that replacement of the blast furnace route by electric steelmaking route is the principal means of achieving that goal. The parties differ on how to realise decarbonisation and that is largely due to the employment consequences of different timings and technological demands.

The HMG/Tata plan lacks in specifics and the earlier-than-planned closure of the coke ovens raises concerns about the plan. Tata will be aware of the technology available and in development to meet its future production plans. It is understandable that it is not Tata that will be producing green electricity, hydrogen or carbon capture and storage systems because these lie with other parties. Whilst the options for Tata in choosing its decarbonisation technology are therefore complex, it could surely outline some of the possible scenarios.

Within that strategy, the Transition Board's organisation and terms of reference are clear, but their acceptance will much depend on the exact process technology, specific timings and manpower requirements in the plan.

The Unite Workers' Plan is clear in terms of the future configuration of the steelworks it proposes. It is also compatible with other decarbonisation plans for south Wales. Unite members tend to have more transferable skills and this may be why its plan recognises the importance of incorporating what is offered by other plans for future employment possibilities.

The GMB/Community/Syndex plan lacks sufficient detail with regards timings and technological possibilities, although it is the only plan that mentions the prospect of using direct reduction of iron ore. Its aim seems to be the retention of the employment for as long as possible of members who possess ironmaking-specific skills. That makes its participation in the People, Skills and Business working group so critical as such specific skills will require transfer training to enable them to do new work.

Tata's plans for the exact configuration of its plant and output targets were unclear.

Table 4 (on page 179) therefore provides a speculative scenario. The early provision of this information would surely have bolstered the confidence of staff and the public.

Port Talbot steel – Unite Workers Plan campaigners
(Unite)

Participant	Celtic Freeport	Swansea Bay City Deal	Swansea Bay and west Wales Metro	Workers Plan for Port Talbot and the Syndex plan	Transition Board and its working groups
Neath-Port Talbot Council	Yes	Yes	Yes	Yes	Yes
Swansea Council		Yes	Yes		
Pembrokeshire Council	Yes	Yes	Yes		
ABP Milford Haven	Yes				
Trade Union Representation	Indirectly	Indirectly	Yes, with seven unions	Yes	Yes

Table 1: Participants involved in decarbonisation plans

Feature	Celtic Freeport	Swansea Bay Deal	Swansea Bay and west Wales Metro	Unite Workers' Plan	Community and GMB Unions Syndex Plan
Digital development		Yes			
Domestic decarbonisation and generation		Home retrofit Homes as power stations			
Education and training	Yes	Yes		Yes	
Health and Life Sciences		Yes		Yes	
Offshore wind	Floating offshore wind project (FLOW)			Floating offshore wind	
Onshore wind				Turbine manufacturing	
Solar power				Battery manufacture Solar farms	
Sustainable fuels	Hydrogen	Hydrogen	Hydrogen buses	Green hydrogen Chemical feedstock	
Tidal energy		Marine engineering testing		Yes	
Transport	Port upgrade Low carbon logistics	Decarbonised journeys Hydro vehicles	Modal change and modal integration		
Other	Carbon capture and storage		Low energy cement New industrial zone		

Table 2: Features of the decarbonisation plans

177

Date	HMG/Tata	Unite Workers'	Community/GMB/Syndex
2023			
September	Announcement of decarbonisation plan		
October	(19th) First meeting of Transition Board (TB)		
November	Second meeting of TB (30th)	Multi-union plan proposed	Multi-union plan proposed
2024			
Q1	19 January: Statutory consultation starts 1 February: 3rd TB 27 March: 4th TB following actual closure of coke ovens in March. Redundancy plans	Phase One Set up an independent investigation into the coke ovens	2024–2028 transition The closure of blast furnace 5. Coke ovens to close
Q2	April: 5th TB Substantive discussion. Presentation of local economic plan. Blast furnace 5 to shut in June	Tata ends consultation 25th April	Tata ends consultation 25th April
Q3			
Q4	Blast furnace 4 to shut		
2025	Continuous annealing plant to shut		
2026			
2027	Electric furnace to start	Replace blast furnace 5 by a 3 million tonne per year (mtpa) electric arc furnace, not the proposed 1.5	
2028		Electric furnace to start up	2028–2031 First electric arc furnace installation. Capacity 3 million tonnes a year. Additional electric arc furnace/direct reduction or OSBF* smelting * Open slag bath furnace

178

Date	HMG/Tata	Unite Workers'	Community/GMB/Syndex
2029		Phase Two
Install extra arc furnaces to increase capacity from present 5 to 6–9 mtpa with local energy generation.
Provide green energy for direct reduction.
Introduce advanced scrap processing.
Create a new industrial zone.	
2030			
2031			
2032		Keep blast furnace 4 until its 2034 end-of-life.	2032: Blast furnace 4 to shut. Replace with second EAF.

Table 3: Timeline of events

Typical electric arc furnace capabilities and operating characteristics to achieve an annual output of 3.5 million tonnes using the (assumed) parameters in the table below might be:

Furnace capabilities	Operation
Capacity	420 tonnes
Tap-to-tap time	1 hour
Production pattern	21 x 8-hour shifts per week
Operating availability	50/52 weeks a year
Yield: scrap to molten steel	90%

Annual output possible (52 weeks) = 420 x 21 x 8 x 52 = 3,669,120 tonnes (100% scrap to molten steel yield)
Annual output possible (52 weeks) = 3,669,120 x 90% = 3,302,208 tonnes (90% scrap to molten steel yield)

To achieve 3.5 million tonnes would therefore require either a reduction in tap-to-tap time or an improved yield in increased furnace size because neither operating availability nor production time could be increased.

Table 4: A scenario for a 3 million tonne electric arc furnace

Conclusion

On 22 May, Prime Minister Sunak suddenly announced that a General Election would be held on 4th July. The Labour Party, firm favourites to form the next government on that date, then detailed its plans.

A statutory Industrial Strategy Council would oversee a ten-year infrastructure strategy within which the following funds would be allocated from a £37.3 billion National Wealth Fund.
- £2.5 billion for the steel industry as a whole
- £1.8 billion for ports and supply chains
- £1.5 billion for battery giga factories
- £1.0 billion for carbon capture and storage
- £0.5 billion for green hydrogen manufacture

The £2.5 billion is in addition to the £500 million promised for Port Talbot as part of the HMG/Tata plan.

A new public company called Great British Energy would be created. Its significance can now be considered with the Port Talbot situation in mind. Although the company would be headquartered in Scotland, its aims and formation raised issues like those involved in the phase out of coal production and the use of coking coal in steelmaking.

The proposal not to issue any new oil and gas licenses led to claims that too many jobs were at risk, but Labour said the past mistakes in phasing out coal would not be repeated.

Great British Energy would be run in similar fashion to EDF (Électricité de France). It would be owned by Government and operate in a competitive market. It would not retail energy but would generate power and own, manage, and operate clean power projects alongside private firms. Initial investments would focus on wind and solar projects, including floating offshore wind, hydrogen production and carbon capture and storage.

Its capitalisation of £8.5 billion would be allocated to two funds:
- A Local Power Plan in the form of a fund for local authorities and community groups to finance clean energy projects.
- An investment found for (GBE) to establish itself in the energy market.

Some of the projects proposed in Table 2 would be potentially eligible for funding applications from the Local Power Plan by participants in Table 1. In May, Tata entered an agreement with National Grid for the supply of power to Port Talbot steelworks from 2027.

The Port Talbot situation is not just about Port Talbot. That's because the proposals to achieve decarbonisation there have national implications. One implication is the need for state procurement in partnership with private business that are disadvantaged because of state subsidies (dumping) favouring competitors. The three groups of proposals are compatible and can provide a template for decarbonisation and industrial change elsewhere eg the oil industry in the north-east of Scotland.

The key agent in implementing decarbonisation and industrial change in south-west Wales is likely to be the Welsh Government whose role will be to ensure the UK infrastructure strategy is implemented through the local plans.

There is now a potential opportunity to alleviate many of the negative consequences of the necessary decarbonising of steelmaking.

After the UK General Election on 4 July 2024, within a new economic model for south-west Wales, decarbonisation need not mean further deindustrialisation.

The answer?

Britain never really understood the implications of the loss of its empire, unlike other countries, especially the losers in War, such as Germany and Japan, did. Britain was prepared to tolerate austerity after 1945 to recover from the damage of war and to create the conditions for better times. When those times came in the 1950s and 1960s with the end of food rationing, the supply of better housing and consumer durable goods, the country perhaps became a little too euphoric, even hubristic, about its real position in the world.

In the 1960s and 1970s, there were those who knew what was needed, in both government and opposition parties. Organised restraint was required to develop the country's infrastructure and industries within an agreed renewal plan which would move the country forwards in line with the best elsewhere. No one succeeded, with the result that, in economic terms, the relative inefficiency of British business led to the country substituting home-produced industrial goods with imports.

By the 1980s, the answer was, seemingly, Thatcherism. For those industries regarded as outdated or inefficient in their working practices, such as coal and steel, their aggressively contrived demise was a matter of too far, too fast. Yet, even wholesale privatisation of public utilities and national assets was attractive to investors for a while, with such carrots as the 'right-to-buy' council houses. Another legacy of Thatcherism in the 1980s was the Big Financial Bang. Financial capitalism served the few very well indeed. It was too often more beneficial to asset strip a company than to invest in it. Capital gains were seen as a much easier way of providing for life than earned income. Outsourcing, both foreign and domestic, was yet another technique to enrich, without providing the alleged benefits to the end consumer.

The advent of overseas investment, such as the arrival of Japanese manufacturing to the UK by the likes of Sony and Nissan, were accepted by some as adequate compensation. Soon, there were few organisations in the UK, whether industrial, financial, or service, that were not foreign owned. Companies like Courtaulds and ICI were fragmented or obliterated altogether. Little wonder that people wanted to 'take back control' in 2016. We have never had a national plan for industry.

It took many a little longer to realise that both privatisation of public services, and later from 2010, the new version of austerity did not work. Public services struggle to exist today. Although many of the developments in the field of information and communication technology are being applied within much of the foreign-owned industrial sector, other innovation is also being applied to sustain the insecurity and inequality of the Gig economy.

This paper just answers how, over the last fifty or sixty years, we have reached the position we are now in. We are once more in an uncertain and unequal world; if we are all to be secure, let alone prosperous, then the explanation and lessons offered here needs to be applied in Britain as soon as possible.

Acronyms and abbreviations

ABP	Associated British Ports
AEEU, AUEF, AUEW	Names of engineering unions before 1992
Amicus	A trade union which merged with the TGWU to form Unite
BSC	British Steel Corporation
CAB	Citizens Advice Bureau
CIA	Central Intelligence Agency
DVLA	Driver and Vehicle Licensing Agency
E.On	A German multinational electricity company
EPA	US Environmental Protection Agency
EV	Electric vehicle
GKN	British multinational automotive and aerospace components business
GMB	General trade union in the United Kingdom for most industrial sectors
GPMU	The Graphical, Paper and Media Union between 1991 and 2005
GRE/AXA	Guardian Royal Exchange Insurance, acquired by AXA of France in 1999
HMG	His/Her Majesty's Government
HSE	Health and Safety Executive
IEMA	Institute of Environmental Management and Assessment
IOSH	Institution of Occupational Safety and Health
ISITB	Iron and Steel industry Training Board
Ludlow 21	A voluntary group to promote sustainable living in the Ludlow area
LSTG	Ludlow Sustainable Transport Group, part of Ludlow 21
MOD	Ministry of Defence
NORCAT	Northumberland College, part of Education Partnership North-East
PPM	Parts per million
RVI	Royal Victoria Infirmary in Newcastle upon Tyne
SNCF	France's national state-owned railway company
TfW	Transport for Wales
TGWU	Transport and General Workers' Union, now part of Unite
UNIFI	A trade union representing finance workers, now part of Unite.
Unite	A trade union formed by the merger of Amicus and the TGWU in 2007